Johannes Baumgart

The Hair Bundle

Johannes Baumgart

The Hair Bundle

Fluid-Structure Interaction in the Inner Ear

Südwestdeutscher Verlag für Hochschulschriften

Impressum/Imprint (nur für Deutschland/only for Germany)
Bibliografische Information der Deutschen Nationalbibliothek: Die Deutsche Nationalbibliothek verzeichnet diese Publikation in der Deutschen Nationalbibliografie; detaillierte bibliografische Daten sind im Internet über http://dnb.d-nb.de abrufbar.
Alle in diesem Buch genannten Marken und Produktnamen unterliegen warenzeichen-, marken- oder patentrechtlichem Schutz bzw. sind Warenzeichen oder eingetragene Warenzeichen der jeweiligen Inhaber. Die Wiedergabe von Marken, Produktnamen, Gebrauchsnamen, Handelsnamen, Warenbezeichnungen u.s.w. in diesem Werk berechtigt auch ohne besondere Kennzeichnung nicht zu der Annahme, dass solche Namen im Sinne der Warenzeichen- und Markenschutzgesetzgebung als frei zu betrachten wären und daher von jedermann benutzt werden dürften.

Coverbild: www.ingimage.com

Verlag: Südwestdeutscher Verlag für Hochschulschriften GmbH & Co. KG
Dudweiler Landstr. 99, 66123 Saarbrücken, Deutschland
Telefon +49 681 37 20 271-1, Telefax +49 681 37 20 271-0
Email: info@svh-verlag.de

Approved by: Dresden, Technische Universität Dresden, Dissertation, 2010

Herstellung in Deutschland:
Schaltungsdienst Lange o.H.G., Berlin
Books on Demand GmbH, Norderstedt
Reha GmbH, Saarbrücken
Amazon Distribution GmbH, Leipzig
ISBN: 978-3-8381-2727-9

Imprint (only for USA, GB)
Bibliographic information published by the Deutsche Nationalbibliothek: The Deutsche Nationalbibliothek lists this publication in the Deutsche Nationalbibliografie; detailed bibliographic data are available in the Internet at http://dnb.d-nb.de.
Any brand names and product names mentioned in this book are subject to trademark, brand or patent protection and are trademarks or registered trademarks of their respective holders. The use of brand names, product names, common names, trade names, product descriptions etc. even without a particular marking in this works is in no way to be construed to mean that such names may be regarded as unrestricted in respect of trademark and brand protection legislation and could thus be used by anyone.

Cover image: www.ingimage.com

Publisher: Südwestdeutscher Verlag für Hochschulschriften GmbH & Co. KG
Dudweiler Landstr. 99, 66123 Saarbrücken, Germany
Phone +49 681 37 20 271-1, Fax +49 681 37 20 271-0
Email: info@svh-verlag.de

Printed in the U.S.A.
Printed in the U.K. by (see last page)
ISBN: 978-3-8381-2727-9

Copyright © 2011 by the author and Südwestdeutscher Verlag für Hochschulschriften GmbH & Co. KG and licensors
All rights reserved. Saarbrücken 2011

Sea Ranch, California, 2008

Lass das Auge die Klänge fangen;
Dann wirst du endlich verstehen.
Tozan, 9. Jahrhundert, Japan

Let your eye catch the sound;
Then you will finally understand.
Tozan, 9th century, Japan

Kurzfassung

Das Haarbündel: Fluid-Struktur-Wechselwirkung im Innenohr

Bei der Hörwahrnehmung eines Klangs spielen viele komplexe Prozesse zusammen. Der Schlüsselprozess, die Umwandlung mechanischer Schwingungsbewegung in elektrische Signale, findet in den Haarbündeln im Innenohr statt. Diese Haarbündel sind hoch entwickelte mechanosensitive Organellen, bestehend aus vielen nahe beieinander stehenden Stereozilien umgeben von Flüssigkeit. Die beträchtliche Viskosität dieser Flüssigkeit führt zur Energiedissipation und zur Schwingungsdämpfung, was im Gegensatz zur bekannten hohen Empfindlichkeit und der ausgezeichneten Frequenzselektivität der Hörwahrnehmung steht. Um die Komponenten des Haarbündelsystems in ihrem funktionalen Zusammenspiel besser zu verstehen, bedarf es eines wirklichkeitsgetreuen Modells unter Einbeziehung der Wechselwirkung zwischen Flüssigkeit und Struktur.

Mit dieser Arbeit wird ein neuer Ansatz vorgestellt, um die Mechanik der Fluid-Struktur-Wechselwirkung im Innenohr zu analysieren. Da die Bewegungen bei der normalen Mechanotransduktion wesentlich kleiner als die geometrischen Abmessungen sind, ist es möglich, das Verhalten von Fluid und Struktur in Form der Verschiebungsvariable in einem linearen einheitlichen System von Gleichungen ausreichend genau zu beschreiben. Dieses System von partiellen Differentialgleichungen wird mit der Finite-Elemente-Methode gelöst. Basierend auf experimentell ermittelten Daten vom Haarbündel des Ochsenfrosches wird ein detailliertes Modell erstellt, welches sowohl die Interaktion mit der umgebenden Flüssigkeit als auch die koppelnde Flüssigkeit in den engen Spalten zwischen den einzelnen Stereozilien erfasst. Die experimentellen Daten sind Ergebnisse von hochauflösenden interferometrischen Messungen bei physiologisch relevanten Bewegungsamplituden im Bereich von unter einem Nanometer bis zu mehreren Dutzend Nanometern, sowie über einen breiten Frequenzbereich von einem Millihertz bis hundert Kilohertz.

Das Modell erlaubt die Berechnung der auftretenden viskosen Widerstände aus der numerischen Analyse der verschiedenen beobachteten Bewegungsmoden. Es kann gezeigt werden, dass durch die Gruppierung zu einem Bündel der Gesamtwiderstand drastisch reduziert ist, im Vergleich zur Summe der Widerstände einzelner Stereozilien, die sich individuell und unabhängig voneinander bewegen. Die einzelnen Stereozilien in einem Haarbündel sind durch elastische Strukturen mechanisch miteinander verbunden: Die Energie des Schalls wird durch schräg angeordnete sogenannte Tiplinks auf die mechanotransduktiven Kanäle übertragen, wohingegen horizontale Querverbindungen die Stereozilien direkt koppeln. Während der Haarbündelauslenkung verursachen die Tiplinks zusätzlichen Widerstand durch stark dissipative Relativbewegungen zwischen den Stereozilien. Die horizontalen Querverbindungen unterdrücken diese Bewegungen und sind dafür verantwortlich, dass sich das Haarbündel als Einheit bewegt und der Gesamtwiderstand gering bleibt. Die Steifigkeit der Stereozilien und der Verbindungselemente sowie deren Geometrie sind in dem Modell sorgfältig angepasst, um eine Übereinstimmung mit den Beobachtungen aus verschiedenen Experimenten zu erzielen. Als Referenz dienen Steifigkeits- und Widerstandsmessungen, sowie Kohärenzmessungen für die gegenüberliegenden Außenkanten des Bündels, die jeweils mit und ohne Tiplinks durchgeführt wurden. Darüberhinaus sind die Ergebnisse durch den Vergleich mit experimentell beobachteten Relativbewegungen validiert, die das Haarbündel infolge von sinusförmiger Anregung bei Distorsionsfrequenzen zeigt. Diese haben ihren Ursprung in dem nichtlinearen Prozess des Öffnens von Ionenkanälen.

Das entwickelte Modell eines Haarbündels liefert neue Einblicke in den Schlüsselprozess der auditiven Wahrnehmung. Zur Behandlung von Problemen der Fluid-Struktur-Wechselwirkungen bei kleinen Amplituden hat sich der hier ausgearbeitete Ansatz als effizient und zuverlässig erwiesen.

Abstract

The Hair Bundle: Fluid-Structure Interaction in the Inner Ear

A multitude of processes cooperate to produce the sensation of sound. The key initial step, the transformation from mechanical motion into an electrical signal, takes place in highly specialized mechanosensitive organelles that are called hair bundles due to their characteristic appearance. Each hair bundle comprises many apposed cylindrical stereocilia that are located in a liquid-filled compartment of the inner ear. The viscous liquid surrounding the hair bundle dissipates energy and dampens oscillations, which poses a fundamental physical challenge to the high sensitivity and sharp frequency selectivity of hearing. To understand the structure-function relationship in this complex system, a realistic physical model of the hair bundle with an appropriate representation of the fluid-structure interactions is needed to identify the relevant physical effects.

In this work a novel approach is introduced to analyze the mechanics of the fluid-structure interaction problem in the inner ear. Because the motions during normal mechanotransduction are much smaller than the geometrical scales, a unified linear system of equations describes with sufficient accuracy the behavior of the liquid and solid in terms of a displacement variable. The finite-element method is employed to solve this system of partial differential equations. Based on data from the hair bundle of the bullfrog's sacculus, a detailed model is constructed that resolves simultaneously the interaction with the surrounding liquid as well as the coupling liquid in the narrow gaps between the individual stereocilia. The experimental data are from high-resolution interferometric measurements at physiologically relevant amplitudes in the range from a fraction of a nanometer to several tens of nanometers and over a broad range of frequencies from one millihertz to hundred kilohertz.

Different modes of motion are analyzed and their induced viscous drag is calculated. The investigation reveals that grouping stereocilia in a bundle dramatically reduces the total drag as compared to the sum of the drags on individual stereocilia moving in isolation. The stereocilia in a hair bundle are interconnected by oblique tip links that transmit the energy in a sound to the mechanotransduction channels and by horizontal top connectors that provide elastic coupling between adjacent stereocilia. During hair-bundle deflections, the tip links induce additional drag by causing small but very dissipative relative motions between stereocilia; this effect is offset by the horizontal top connectors that restrain such relative movements, assuring that the hair bundle moves as a unit and keeping the total drag low. In the model the stiffness of the links, the stiffness of the stereocilia, and the geometry are carefully adjusted to match experimental observations. The references are stiffness and drag measurements, as well as the coherence measurements for the bundle's opposite edges, both with and without the tip links. The results are further validated by a comparison with the relative motions measured in a sinusoidally stimulated bundle for the distortion frequencies at which movements are induced by the nonlinearity imposed by channel gating.

The model of the fluid-structure interactions described here provides insight into the key step in the perception of sound and the method presented provides an efficient and reliable approach to fluid-structure interaction problems at small amplitudes.

Contents

1 Introduction   1
     1.1 Objectives and Motivation 1
     1.2 State of the Art 4
     1.3 Overview 7

2 Equations   9
     2.1 Governing Equations 9
         2.1.1 Conservation Equations 10
         2.1.2 Kinematics 11
         2.1.3 Material Description 11
         2.1.4 Equations of Motion for a Newtonian Fluid 14
     2.2 Material Properties 15
         2.2.1 Properties of the Hair Bundle Materials 15
         2.2.2 Properties of the Liquids in the Inner Ear 17
         2.2.3 Interaction of Liquid and Solid 19
     2.3 Simplified Equations for Inner-Ear Mechanics 19
         2.3.1 Dimensionless Form of Fluid Equations 20
         2.3.2 Small-Amplitude Fluid Equations 25
         2.3.3 Small-Amplitude Solid Equations 26
         2.3.4 Coupled Fluid and Solid Equations for Small Amplitudes . . . 27

3 Finite-Element Method   29
     3.1 Finite-Element Discretization 29
         3.1.1 Galerkin Projection 29
         3.1.2 Frequency Formulation 34
         3.1.3 Implementation 35
     3.2 Validation of the Finite-Element Formulation 35
         3.2.1 Oscillating Sphere 35
         3.2.2 Acoustic Tube 37
         3.2.3 Fluid-Filled Elastic Sphere 40

4	**Model of the Hair Bundle**		**43**
	4.1	Geometry	43
	4.2	Mechanical Properties	50
	4.3	Finite-Element Mesh	52
	4.4	Material Properties	56
5	**Bundle Mechanics**		**59**
	5.1	Kinociliary Bulb Displaced	60
		5.1.1 Drag	63
		5.1.2 Stiffness	70
		5.1.3 Inertia	74
	5.2	Coupling between Stereocilia	75
	5.3	Response to Tip-Link Forces	79
	5.4	Analysis of Relative Motions in the Bundle	82
6	**Conclusion**		**91**
	6.1	Summary	91
	6.2	Directions for Future Research	93
	6.3	Acknowledgments	94

Bibliography	**97**
List of Figures	**105**
List of Tables	**107**
List of Symbols	**109**

A	**Units**		**115**
B	**Two Cylinder Drag**		**117**
	B.1	Pivoting to their Common Center	117
	B.2	Pivoting in the Same Direction	119

C Parameters 121
C.1 Wall-to-Wall Distance at the Tips 122
C.2 Tip-Link Stiffness 123
C.3 Top-Connector Stiffness 124
C.4 Kinocilium Stiffness 125
C.5 Pivotal Stereocilia Stiffness 126
C.6 Viscosity between Stereocilia 127
C.7 Density 128

Chapter 1

Introduction

1.1 Objectives and Motivation

Our hearing organ is able to detect air vibrations with an amplitude smaller than the Bohr radius while cycling about a thousand times per second.* These faint amplitudes are smaller than the thermal motion of the mechanosensitive structures. Besides this remarkable ability to detect tiny amplitudes, hearing differentiates minuscule variations of frequency modulation and of spectral content.† To gain further insights into the mechanical processes underlying the sensation of sound motivates this work.

A key step in the perception of sound is the transformation from mechanical motion into an electrical signal, termed mechanotransduction. In all vertebrates this takes place at the hair bundles which are situated atop hair cells. These cells occur in the labyrinth of the inner ear and in the lateral lines of fishes and aquatic amphibians.

This work analyzes the fluid-structure interaction of the hair bundle in the inner ear by means of numerical modeling. The inner ear is filled with an aqueous solution that on a micrometer geometrical scale and at kilohertz frequencies is highly viscous and nearly incompressible.

Scaling the geometry of the bundle up to the size of a human hand illustrates the environment that the bundle experiences. The hand is about 10,000 times larger than a hair bundle. For the hair bundle oscillatory motions with about 1,000 cycles per second are characteristic, but for the hand the motions are just in the order of one per second. For self-similar behavior the ratio of inertial to viscous force must be preserved, which implies that the length scale squared multiplied by the time scale has to be constant. One way to ensure the same force ratio is by adjusting the properties of the surrounding liquid. To resemble a hair bundle, the hand would have to move through a liquid with the dynamic viscosity of honey!

---

*Based on 0 dB SPL hearing threshold at 1 kHz and an impedance of 420 Pa·s·m^{-1} for air. The Bohr radius is about 53 pm and the most likely position of the electron orbiting around the hydrogen atom.

†A healthy person can detect frequency modulations of about 0.1 %.

A hair bundle comprises many closely apposed, cylindrical stereocilia. The stereocilia protrude from the cuticular plate, the upper surface of the hair cell. They almost touch each other at their tips, while standing further apart at their basal insertions. The very small gaps between the stereocilia as well as the volume surrounding the bundle are filled by liquid, the endolymph. The bundle has a distinct direction parallel to the cuticular plate: the excitatory direction. When a bundle is displaced along this direction, mechanically sensitive ion channels open and an ion flow depolarizes the cell. In the positive direction the stereocilia increase in height. In this direction each shorter stereocilium is also linked by an oblique tip link to the adjacent taller stereocilium. The stereocilia consist mainly of actin filaments cross-linked by fimbrin, espin, and perhaps other proteins.

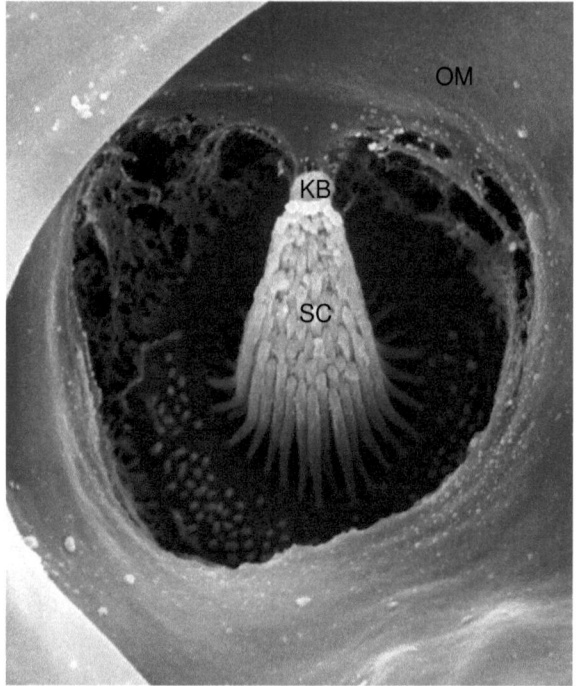

Figure 1.1: Scanning electron micrograph of a hair bundle from the bullfrog's sacculus attached to the otolithic membrane. The kinocilium with the kinociliary bulb resides at the bundle's tall edge. The stereocilia form a hexagonal pattern. The calcium carbonate crystals above the bundle are removed. The excitatory direction is toward the upper right. KB: Kinociliary bulb, SC: Stereocilia, OM: Otolithic membrane. *Courtesy of Dr. A. J. Hudspeth, Howard Hughes Medical Institute*

1.1 Objectives and Motivation

As an example, the geometry of the hair bundle of the bullfrog sacculus is given in detail (Figure 1.1). In a top view the stereocilia form a hexagonal pattern. The region where the stereocilia connect to the cuticular plate has a circular shape. The kinocilium resides at the tall edge of the bundle and has a similar shape as the stereocilia. The kinociliary bulb at the tip of the kinocilium couples the bundle to the otolithic membrane. In this work the hair bundle of the bullfrog sacculus is used. Although the human hearing organ senses higher frequencies, the bullfrog hair bundle provides some experimental advantages. The characteristic frequencies are lower and the preparation is usually more stable and viable *ex vivo*. The large distance between the tall and short edge of the bundle allows one to clearly distinguish them with optical measurement techniques.

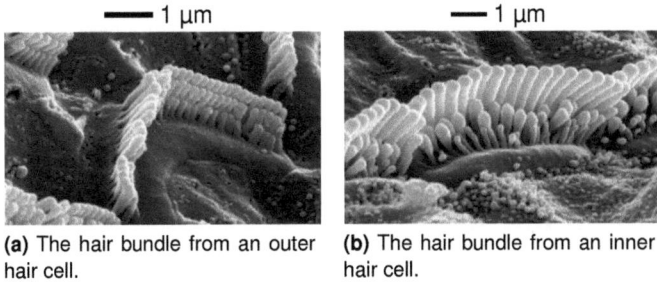

(a) The hair bundle from an outer hair cell.

(b) The hair bundle from an inner hair cell.

Figure 1.2: Scanning electron micrographs of the hair bundles from hair cells of a guinea pig. The overlaying tectorial membrane is removed. The excitatory direction is upwards to the left. *Courtesy of Dr. Y. M. Yarin, Technische Universität Dresden*

The hair bundles of the mammalian and human inner ear are not grouped into a circular arrangement; instead the stereocilia line up in three rows in the bundle. The stereociliary rows of the outer hair cell bundles form a shape like the letters V or W. Here, the height decrease from row to row is more pronounced and there are typically just three rows in a column. The micrographs in Figure 1.2 show hair bundles of a guinea pig from the outer and inner hair cells. While the outer hair cell bundles kink, those of the inner hair cells feature straight rows. Between the individual outer hair cell bundles is a gap in the longitudinal direction, whereas the edges of the outer hair cells almost touch each other. The mature mammalian hair bundles lack kinocilia. The tallest row of the outer hair cell bundle is mechanically connected to the overlaying tectorial membrane. The hair bundles of inner hair cells are free-standing.

To accomplish the mechanical analysis of the elastic hair bundle in the liquid environment, appropriate numerical tools are necessary. Because an elastic solid and a viscous fluid are present and coupled, this poses a problem of fluid-structure interaction. The behavior of the physical fields is described by casting them into differential equations on the basis of mass, momentum, and energy conservation.

The mechanics of hearing are characterized by micrometer dimensions, nanometer displacements, and kilohertz frequencies. These special conditions allow us to neglect the large-deformation description for solid material and convection for fluid. The remaining system of coupled differential equations is linear. To solve the equation on the complex, three-dimensional geometry of the hair bundle, the finite-element method is employed. A novel approach is presented to efficiently solve the strongly coupled two-field problem. The implementation is successfully tested with problems related to hearing mechanics.

For this work extensive experimental data and geometrical observations where provided by Dr. A. J. Hudspeth and Dr. Andrei S. Kozlov,* Laboratory of Sensory Neuroscience, Howard Hughes Medical Institute, Rockefeller University, New York. Support on the theoretical aspects and data analysis of hair bundle data was given by Dr. Thomas Risler, Institut Curie, Physical Chemistry, Paris.

1.2 State of the Art

For several decades the field of inner ear mechanics has been attracting researchers to investigate the mechanism of sound sensation. Already two decades ago important features were identified as reviewed by Hudspeth (1989). The complex nature of hearing involves people from diverse fields and with multiple approaches and interests. More than a century ago von Helmholtz (1896) analyzed the capabilities of the ear with respect to the high precision when listening to music. Later von Békésy (1960) revealed the underlying mechanism to differentiate frequencies motivated by the research on telecommunication signal quality for the Hungarian Post Office. Using cadavers, he experimentally showed how the traveling wave transports the mechanical energy to different locations depending on the stimulus frequency. Later experimental evidence was found that the ear is not solely a passive sensor, but is active and produces sounds, which is nowadays used in clinics to test the hearing of newborns. The idea of an active process was originally proposed by Gold (1948) and later experimental evidence was found at the level of individual hair cells by Ricci et al. (2000) and by Martin et al. (2001). There are several signatures for this active process, such as amplification, frequency tuning, compressive nonlinearity, and spontaneous otoacoustic emissions (Hudspeth 2008; Robles and Ruggero 2001). Today most of the involved individual components are characterized in their mechanical properties (Puria and Steele 2008), but their interplay is still discussed (Ashmore et al. 2010).

*Current address: Department of Psychology, University of California, San Diego

1.2 State of the Art

The key step in the process of hearing is the transformation from the mechanical stimulus to an electrical signal. The hair bundle in the inner ear is sensitive to motion as channels open while displaced (Gillespie and Müller 2009). The mechanical relations were identified by Howard and Hudspeth (1988) and Howard et al. (1988). The intrinsic noise and the specific properties of the hair bundle system might explain the compressive non-linearity as well as frequency tuning (Nadrowski et al. 2004; Hudspeth et al. 2010). The latter is further enhanced by elastic coupling between hair bundles (Dierkes et al. 2008).

To investigate, in which mechanical environment the bundle is working and what are the governing forces in a hair bundle, several numerical models have been developed. Geisler (1993) investigated the strain of the elastic links in the bundle during deflection. Also Pickles (1993) found a linear relationship between the bundle displacement and the tip-link strain. The dependence of the mechanical load on the tip links with respect to the geometrical parameters was analyzed by Cotton and Grant (2004). Using the finite-element method Silber et al. (2004) studied the elastic mechanical deformation on utricular hair bundles. Nam et al. (2007) computed the response of three-dimensional hair bundles with non-linear gating springs to account for the channel gating. To the author's knowledge, the only work incorporating viscous effects in the hair bundle is by Zetes and Steele (1997). Therein the fluid forces between individual stereocilia were estimated based on squeeze-flow assumptions and modeled in the whole bundle arrangement. The interaction of the bundle with the surrounding liquid was analyzed by Freeman and Weiss (1990) and later in more detail by Shatz (2004). Nam et al. (2005) modeled the fluid induced displacement of the hair bundle by applying drag force estimates of the free standing parts of the stereocilia to a three-dimensional finite-element model. Matsui et al. (2006) used as well an analytical estimation for the liquid forces acting on the hair bundle to compute the tension in the links by means of a finite-element model.

Also for the whole organ of Corti, where the hair bundle and cell are placed in, different models and approaches exist. Andoh and Wada (2004) focused with a fluid-structure interaction model on the motion pattern for the fast and slow wave in the organ of Corti. Further on Andoh et al. (2005) computed the phase relation of the inner hair cell bundle displaced by the circumfluent liquid. The employed fluid-structure interaction method is based on an iterative approach in the time domain. Cai et al. (2004) identified a set of material properties for the organ of Corti based on experimental observations and analyzed the motion pattern of the organ of Corti in the two-dimensional cross-section, without viscosity in the liquid. Steele et al. (2009) included the effect of the viscous traveling wave by a Wentzel-Kramers-Brillouin approximation and estimated local drag losses in the organ by analytical estimations. Ramamoorthy et al. (2007) modeled the fully coupled mechano-electro-acoustical system of the cochlea with lumped damping properties accounting for viscous losses adjusted to match the overall response to experimental observations.

So far an efficient approach is missing to compute the strongly coupled fluid-structure interaction problem on the complex geometry of the inner ear. In classical engineering problems the fluid-structure interaction problem is solved by coupling a solver for the fluid mechanics based on the variables velocity and pressure with a solver for the solid mechanics based on the displacements by an iterative procedure. At the interface of solid and fluid the continuity of momentum and displacement has to be ensured. As the fluid and solid domains are solved separately, several iteration steps are necessary for each time step. A way to circumvent this difficulty is to use a single set of conservation equations for the whole domain and account for the different materials by different material laws. Greenshields and Weller (2005) presented this technique, grounded in the fluid mechanical conservation equations with velocity variables. Also Papadakis (2008) used this approach based on velocities with the advantage that the continuity of momentum and displacement is satisfied in each iteration step. To the authors knowledge no unified approach was presented so far for the variables pressure and displacement, which turns out to be an useful approach for the small amplitude motions in hearing mechanics.

A further motivation for this work is the experimental observation by Kozlov et al. (2007). They observed a coherent motion of the hair bundle over broad frequency ranges. The open question is, which forces are keeping the bundle together. A possible answer is the liquid-mediated coupling (Kozlov et al. 2009; Baumgart et al. 2009b; 2010). Based on further experimental data (Kozlov et al. 2011) this work investigates the viscous liquid forces and elastic coupling by the links between stereocilia in detail.

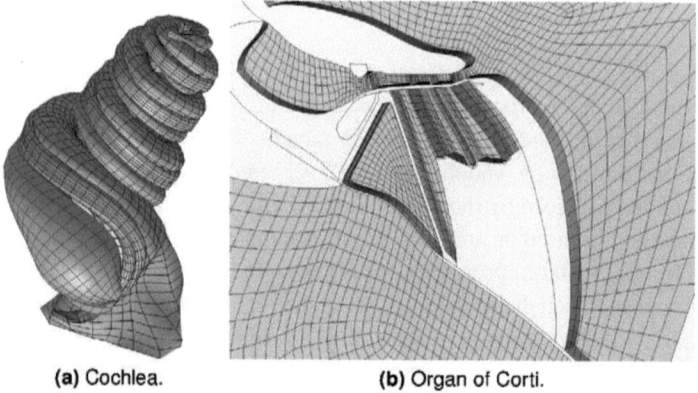

(a) Cochlea.   (b) Organ of Corti.

Figure 1.3: Models of the guinea pig inner ear. *Preliminary work*

Preliminary work was conducted in a collaborative research project on *models of the fluid-structure interaction for the mechanics and pathomechanics of the inner ear* (Fluid-Struktur-Modelle zur Mechanik und Pathomechanik des Innenohres, supported by DFG). The focus of the subproject *macro and micro fluid mechanics (computational-fluid mechanics) in the cochlea* (Makro- und Mikrofluidmechanik (CFD) der Cochlea, GR 1388/14) was on the fluid mechanics in the organ of Corti. At first, the characteristics of the fluid mechanics were analyzed (Baumgart et al. 2007). The geometrical data were provided by micrographs and tomographic scans (Poznyakovskiy et al. 2008) to accomplish realistic models and dynamical measurements were the reference to adjust the properties (Nowotny and Gummer 2006; Chiaradia et al. 2009). As an example the surface reconstruction of the cochlea is presented in Figure 1.3a. Based on the detailed geometry a finite-element model of the organ of Corti of the guinea pig was constructed (Baumgart et al. 2009a). The model is three-dimensional with a width of half a hair cell (Figure 1.3b). The fluid-structure interaction was implemented by describing the fluid mechanics in the context of solid mechanics. The displacement field in the subtectorial space is fully resolved and relations of the deflection of the hair bundles from the inner hair cells to variations of the geometry were investigated. This approach for the modeling of a problem of hearing mechanics was further developed in this work and applied to the hair bundle from the bullfrog's sacculus.

1.3 Overview

In Chapter 2, a set of differential equations is derived. The description is formulated by the displacement field for the solid and fluid material, which adequately represents the small amplitude of motion characteristic of inner-ear mechanics. The pressure-displacement relation of the fluid adds the pressure variable for the fluid. The equations for the fluid originate from the Navier-Stokes equation and are simplified based on dimensionless number analysis.

In Chapter 3, the simplified equations of the fluid are discretized by the finite-element method in a displacement-pressure formulation. Three examples validate the implementation: an oscillating sphere, the impedance in a tube, and a fluid sphere surrounded by an elastic shell.

In Chapter 4, a three-dimensional geometrical model of the hair bundle of the bullfrog's sacculus is constructed based on literature data and micrographs. This model is meshed by finite elements such that the fluid coupling between individual stereocilia as well as the drag by the external liquid are captured.

In Chapter 5, a comprehensive analysis with the finite-element model of the hair bundle provides detailed insights into the mechanical coupling. Three different load cases are investigated: the bundle's response to stimulation at the kinociliary bulb, the place where physiological excitation occurs; mechanical coupling in the bundle between the stereocilia; and the response of the bundle to a force along the tip links.

Chapter 2

Equations for the Liquids and Solids in the Inner Ear

Describing the mechanics of inner-ear materials by differential equations provides a basis to understand the mechanisms of hearing. As the geometrical dimensions measure a multiple of the molecules, the underlying continuum assumption is valid. The focus is on small amplitudes related to the threshold of hearing to gain insights into the function of the highly sensible hearing organ. Dimensionless numbers provide the possibility of simplifying the field equations for the fluid. For smallest oscillatory motions they simplify to a linear set of equations and can be written in terms of displacement. Also the solid material is described by linear differential equations in terms of the displacement. This allows us to describe the solid and fluid by the same set of variables and differential equations. Only the material law differs. For the fluid the two material properties, viscosity and bulk modulus, differ by several orders of magnitude for the frequency range of interest. To circumvent order of magnitude errors for the fluid the additional variable pressure is introduced by a material law and coupled to the displacement.

2.1 Governing Equations

The principle of conservation of mass, momentum, and energy is applied to the fluid and solid materials. In the differential description the kinematic relations link the macroscopically observed displacements and velocities to local strains and strain rates. Phenomenological observations of the material behavior complete the problem. This provides a set of differential equations for the field of observed variables.

The fluid is able to move over large distances compared to the circumfluent boundary. To deal with this large motion it is common to write the equations in the Eulerian frame of reference. Hereby a fluid particle is observed as it moves through a fixed reference frame. This is in contrast to the Lagrange description where the fluid particle is monitored as moving with respect to the environment, which is the classical view in solid mechanics. The mechanics are investigated from the perspective of fluid mechanics and hence an Eulerian description is appropriate here.

2.1.1 Conservation Equations

A fluid as well as a solid material behavior is characterized by the conservation of mass, momentum, and energy. These equations are briefly summarized. More details about their derivation are available in textbooks. The equations are in line with the work of Anderson et al. (2009).

The physical principle of mass conservation can be written as continuity equation

$$\frac{D\varrho}{Dt} + \varrho \nabla \cdot \vec{v} = 0 \tag{2.1}$$

for the time t, density ϱ, and velocity $\vec{v}$. Here the substantial derivative

$$\frac{D\varrho}{Dt} \equiv \frac{\partial \varrho}{\partial t} + \vec{v} \cdot \nabla \varrho$$

combines the local with the convective derivative. Newton's second law provides the basis for the momentum equation:

$$\varrho \frac{D\vec{v}}{Dt} = \nabla \cdot \boldsymbol{\sigma} + \varrho \vec{f} \tag{2.2}$$

for the stress tensor $\boldsymbol{\sigma}$. The external local force per mass $\vec{f}$ can incorporate for example gravity effects. The substantial derivative

$$\frac{D\vec{v}}{Dt} \equiv \frac{\partial \vec{v}}{\partial t} + (\vec{v} \cdot \nabla) \vec{v}$$

sums the local and the convective derivative for the velocity vector. From the angular momentum

$$\boldsymbol{\sigma} = \boldsymbol{\sigma}^\mathsf{T} \tag{2.3}$$

the symmetry of the stress tensor is required. The conservation of total energy reads

$$\varrho \frac{De_\text{tot}}{Dt} = \nabla \cdot (\boldsymbol{\sigma} \cdot \vec{v}) - \nabla \cdot \vec{q} + \varrho \dot{q} + \varrho \vec{f} \cdot \vec{v} \tag{2.4}$$

with $\vec{q}$ the heat flux and $\dot{q}$ the external heat per mass. The total derivative is analogous to the definition of the scalar continuity equation (2.1). The total energy e_tot

$$e_\text{tot} = e_\text{in} + \frac{1}{2} \vec{v} \cdot \vec{v}$$

consists of the internal energy e_in and the kinetic energy.

2.1.2 Kinematics

The strains are a function of the local gradient of the displacement $\vec{u}$. For infinitesimal strains, where large deformations are neglected, this simplifies to

$$\boldsymbol{\varepsilon} = \frac{1}{2}\left(\nabla \vec{u} + (\nabla \vec{u})^\mathsf{T}\right) \tag{2.5}$$

and similarly for the strain rate

$$\dot{\boldsymbol{\varepsilon}} = \frac{1}{2}\left(\nabla \vec{v} + (\nabla \vec{v})^\mathsf{T}\right) \tag{2.6}$$

as a function of the gradient of the velocity $\vec{v}$. Here the velocity

$$\vec{v} = \frac{\partial \vec{u}}{\partial t} \tag{2.7}$$

is the partial time derivative of the displacement $\vec{u}$ for small amplitudes of motion.

2.1.3 Material Description

The conservation equations of mass (2.1), momentum (2.2), and energy (2.4) in the written form contain more variables than there are independent equations. To complete the problem a description of the material is necessary.

The derivation of the equations for the inner-ear mechanics is restricted to linear isotropic materials. Justifications are given in Section 2.2. An isotropic material has no directional dependence of the material law, that relates stresses and strains. Furthermore the linear material is not dependent on the amount of deformation. Based on these assumptions it is sufficient to describe the material behavior for infinitesimal small stresses, strains, and strain rates.

For an isotropic and linear Voigt material the relation between the tensors of stress $\boldsymbol{\sigma}$, strain $\boldsymbol{\varepsilon}$, and strain rate $\dot{\boldsymbol{\varepsilon}}$ is

$$\boldsymbol{\sigma} = 2\mu\boldsymbol{\varepsilon} + \lambda\operatorname{tr}(\boldsymbol{\varepsilon})\boldsymbol{I} + 2\mu'\dot{\boldsymbol{\varepsilon}} + \lambda'\operatorname{tr}(\dot{\boldsymbol{\varepsilon}})\boldsymbol{I} \tag{2.8}$$

with the trace function $\operatorname{tr}(\cdot)$ and $\boldsymbol{I}$ the second-order identity tensor. The four scalar coefficients μ, λ, μ', and λ' characterize the material and are constant.

The stress tensor can be separated into

$$\boldsymbol{\sigma} = -p\boldsymbol{I} + \boldsymbol{\tau} \tag{2.9}$$

the hydrostatic pressure p and the deviatoric stress tensor $\boldsymbol{\tau}$. A fluid at rest exhibits no deviatoric stresses

$$\operatorname{tr}(\boldsymbol{\tau}) = 0 \,. \tag{2.10}$$

The pressure

$$-p = \lambda_{\text{fluid}} \, \text{tr}(\boldsymbol{\varepsilon}) \tag{2.11}$$

is proportional to the negative volume strain. The linear coefficient is labeled for fluids as bulk modulus

$$K = \lambda_{\text{fluid}} \, . \tag{2.12}$$

From thermodynamics this coefficient is

$$K = \varrho \left. \frac{\partial p}{\partial \varrho} \right|_s \tag{2.13}$$

for a process at constant entropy s. The speed of sound is given by

$$c = \sqrt{\frac{K}{\varrho}} \, . \tag{2.14}$$

The coefficient of the other static term μ in the isotropic material (2.8) has to vanish

$$\mu_{\text{fluid}} = 0 \tag{2.15}$$

to incorporate the property of fluids not to withstand static shear deformation (2.10). After finite time the fluid relaxes.

Newton (1726) observed that the deviatoric stresses for a fluid

$$\boldsymbol{\tau} = 2\mu'_{\text{fluid}} \, \dot{\boldsymbol{\varepsilon}} + \lambda'_{\text{fluid}} \, \text{tr}(\dot{\boldsymbol{\varepsilon}}) \, \boldsymbol{I} \tag{2.16}$$

are proportional only to the strain rates. For further simplifications Stokes (1849) chose

$$\lambda'_{\text{fluid}} = -\frac{2}{3}\mu'_{\text{fluid}} \tag{2.17}$$

and describes hereby the material using the dynamic viscosity

$$\eta = \mu'_{\text{fluid}} \, , \tag{2.18}$$

a single scalar coefficient. Hereby no pressure is induced by shear rates. The remaining variable of an isotropic fluid is commonly named bulk viscosity

$$\eta^B = \frac{2}{3}\mu'_{\text{fluid}} + \lambda'_{\text{fluid}} \, . \tag{2.19}$$

This fluid property plays a minor role in classical fluid mechanical problems (Graves and Argrow 1999). It has an effect on sound propagation over distances large compared to the wavelength. This is not the focus of this work. The material behavior is further

2.1 Governing Equations

simplified by setting

$$\eta^B = 0 . \tag{2.20}$$

To sum up, the material law of an isotropic linear fluid is for the pressure

$$p = -K \operatorname{tr}(\boldsymbol{\varepsilon}) \tag{2.21}$$

and for the deviatoric stresses

$$\boldsymbol{\tau} = \eta \left(2\dot{\boldsymbol{\varepsilon}} - \frac{2}{3} \operatorname{tr}(\dot{\boldsymbol{\varepsilon}}) \, \boldsymbol{I} \right) . \tag{2.22}$$

To link this to the observable quantities displacement and velocity the kinematic relations (2.5) and (2.6) are incorporated as

$$\boldsymbol{\tau} = \eta \left((\nabla \vec{v} + (\nabla \vec{v})^\mathsf{T}) - \frac{2}{3} \nabla \cdot \vec{v} \, \boldsymbol{I} \right) \tag{2.23}$$

$$p = -K \nabla \cdot \vec{u} \tag{2.24}$$

with the material parameters bulk modulus K and dynamic viscosity η.

Besides the material law for the fluid also the relations for the solid material are needed. A purely linear elastic material is assumed for the investigated mechanical problems in a first approximation. The assumption is further that an isotropic behavior adequately approximates the bulk properties of the elastic material. No damping is incorporated and thus strain rate-dependent terms are dropped. The material properties are further discussed in Section 2.2.1 and the amplitudes of motion in Section 2.3.1.

The material law (2.8) reads

$$\boldsymbol{\sigma} = 2 \mu_{\text{solid}} \, \boldsymbol{\varepsilon} + \lambda_{\text{solid}} \operatorname{tr}(\boldsymbol{\varepsilon}) \, \boldsymbol{I} \, , \tag{2.25}$$

where the two Lamé parameters of the solid are linked to the engineering values Young's modulus E and Poisson's ratio ν by

$$\mu_{\text{solid}} = \frac{E}{2(1+\nu)} \tag{2.26}$$

and

$$\lambda_{\text{solid}} = \frac{E \, \nu}{(1+\nu)(1-2\nu)} . \tag{2.27}$$

The kinematic relation (2.5) allows us to describe the linear elastic solid material by

$$\boldsymbol{\sigma} = \mu_{\text{solid}} \left(\nabla \vec{u} + (\nabla \vec{u})^\mathsf{T} \right) + \lambda_{\text{solid}} \nabla \cdot \vec{u} \, \boldsymbol{I} . \tag{2.28}$$

Finally, in both solid and fluid Fourier's law relates the heat flux $\vec{q}$

$$\vec{q} = -k_{\text{th}} \nabla T \qquad (2.29)$$

by the thermal conductivity k_{th} to the gradient of the temperature T. The negative sign accounts for the heat flow from high to low temperature.

2.1.4 Equations of Motion for a Newtonian Fluid

The relations between the different physical variables mentioned in the preceding section can be incorporated into the conservation equations. The mass-conservation equation (2.1) stays untouched.

The material description based on Stokes hypothesis for the fluid (2.23) used in the momentum equations (2.2) without external forces reads

$$\varrho \frac{D\vec{v}}{Dt} = -\nabla p + \nabla \cdot \left(\eta \left(\nabla \vec{v} + (\nabla \vec{v})^{\mathsf{T}} \right) - \frac{2}{3} \eta \nabla \cdot \vec{v} \, \boldsymbol{I} \right). \qquad (2.30)$$

The energy equation is rewritten for the fluid without external sources in terms of the directly measurable temperature T

$$\varrho c_p \frac{DT}{Dt} = \nabla \cdot (k_{\text{th}} \nabla T) + \beta T \frac{Dp}{Dt} + \Phi \qquad (2.31)$$

with the dissipation function

$$\Phi = \nabla \cdot (\boldsymbol{\tau} \cdot \vec{v}) - \vec{v} \cdot \nabla \cdot \boldsymbol{\tau} \qquad (2.32)$$

based on the deviatoric stress tensor given by (2.23).[*] Thermodynamic relations provide the isobaric specific heat capacity

$$c_p = \left. \frac{\partial s}{\partial T} \right|_p \qquad (2.33)$$

and the coefficient of thermal expansion

$$\beta = -\frac{1}{\varrho} \left. \frac{\partial \varrho}{\partial T} \right|_p. \qquad (2.34)$$

The latter can also be expressed as

$$\beta = \sqrt{\frac{K \varrho c_p}{T} \left(\frac{c_p}{c_v} - 1 \right)} \qquad (2.35)$$

---
[*]A detailed derivation can be found in textbooks as for example from Schlichting and Gersten (2000).

with the isochoric specific heat capacity (Gaskell 2008)*

$$c_v = \frac{\partial s}{\partial T}\bigg|_v . \qquad (2.36)$$

2.2 Material Properties

The physical properties of the solids and liquids[†] in the inner ear are discussed in this section. The quantities are given in a system of units suitable for inner-ear mechanics. The length is in micrometers, the mass in nanograms, and the time in milliseconds. Further details are given in the Appendix A.

2.2.1 Properties of the Hair Bundle Materials

To be able to set up a continuum model, experimental observations are necessary. If all boundary conditions are known, the material properties can be identified. The tiny dimensions of the cellular materials complicate the measurement of their mechanical response under defined conditions and in idealized load situations. So far an effective approach is to identify the unknown material properties by matching experimental observations with a model reproducing the geometry as closely as possible and already identified material characteristics.

If the experimental situation coincides with the physiological reality and the model is sufficiently accurate, this technique provides reliable suitable material properties for the model. The estimated properties are not necessarily general such that they can be used for the same cell material in other load configurations or different geometrical arrangements. To investigate physiologically relevant situations with numerical models it is therefore advantageous to have numerical models and experimental situation and load configuration as close as possible to the *in vivo* conditions.

Spatial models of the inner-ear mechanics use orthotropic (Fleischer et al. 2010; Steele et al. 2009) or isotropic (Andoh and Wada 2004; Cai et al. 2004) elastic material laws for the various cell types. The cell with wall and soma is not resolved, instead a homogenous material is used. These models reproduce the geometry very accurately. The material properties are adjusted by various observations. It is remarkable how well these models can predict mechanical responses by this calibration technique.

Typical values for the Young's modulus in spatial models are in the range from a few kilopascals to gigapascals. The upper limit is motivated by the properties of the actin filaments and microtubules in the stiff cytoskeleton (Gittes et al. 1993). A lower limit of the stiffness is not possible to be defined from the ultrastructure. The liquid in the

---

*The relations given and the energy equation are based on the assumption of thermal equilibrium and without chemical reactions.

[†]To emphasize the water characteristics the term liquid is used and not fluid.

cells provides a relevant lower bound as an oscillating object in a liquid environment generates viscous forces. If the force produced by the liquid is larger than that from the elastic interior of the solid body, the elastic properties of the cell do not play a role. In water the equivalent shear modulus at 1 kHz is for example about 1 Pa.

Also three-dimensional macroscopic models of the hair bundle use the approach of matching experimental observations with the numerical model (Cotton and Grant 2004; Nam et al. 2006). If the model to analyze the experimental data agrees with the numerical model, parameters can be used directly, as for example in the model by Nadrowski et al. (2004).

Another approach to obtain material properties is the direct simulation on an atomistic level. So far only very well studied small structures as the tip links (Gillespie and Müller 2009) can be computed. The stiffness of the cadherin complex of the tip links was computed by Sotomayor et al. (2010). The values there are about 50 times higher than the experimental observations of Howard and Hudspeth (1988). Although such an approach might be in general desirable, the current state of the art is far from being usable for large complexes in terms of molecular dimensions as the hair bundle.

In this work the approach of matching the properties to experimental observations was used. New experimental data were available from Kozlov et al. (2007; 2011) and a new model was build, therefore the material parameters had to be redefined. Details are given in Section 4.2.

The cells consist of a stiff skeleton filled up by liquid. In some cells the supporting structure possess a distinct orientation, as for example in the stereocilia (Volkmann et al. 2001) and pillar cells (Tolomeo and Holley 1997). In the stereocilia actin filaments align with the axis and have a hexagonal arrangement in a cross section. The actin filaments are cross-linked by fimbrin, espin, and other proteins (Shin et al. 2007). Consequently an anisotropic material law should be used. For the simplest suitable case of a transversal, isotropic, elastic material six independent material parameters are necessary.* As there are not sufficient independent experiments to unambiguously identify six material parameters, the material is regarded as isotropic. A further simplification is that the material properties are assumed constant in every region. As all structures mainly consist of water, it is assumed that the density of all materials is that of water.

Another approach to identify the unknown parameters of an anisotropic material might be to use homogenization techniques to retrieve macroscopic equivalent properties from the microstructure. To accomplish this, the geometry and mechanical properties of the skeleton are needed. The connection of cross links to the actin filaments is unclear and especially important for the mechanical behavior (Bathe et al. 2008).

---

*For an isotropic material just two independent parameters are needed, as for example the Young's modulus and Poisson's ratio. For an transversal isotropic material the properties are isotropic in a plane. In this plane the Young's and shear moduli and the Poisson's ratio are necessary as well as the Young's modulus for the normal direction and the shear modulus and Poisson's ratio between this direction and the isotropic plane.

2.2.2 Properties of the Liquids in the Inner Ear

There are two liquids in the inner ear, the endolymph around the hair bundle in the scala media and the perilymph in scala vestibuli and tympani. Although the constituents are different, the overall ionic concentration (Jahn and Santos-Sacchi 2001) is about the same of the two liquids. Experimental investigations by Zwicker (1974) on the extracted liquids of pig inner ears confirm a viscosity close to that of seawater. He measured separately the viscosity of perilymph and endolymph with differences between both smaller than the uncertainty. Dahl and Kleinfeldt (1970) were using spheres sinking in the perilymph in the cochlea of the guinea pig.

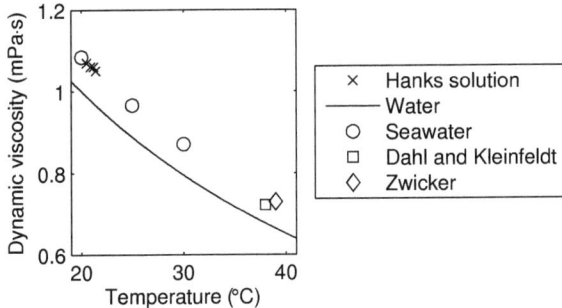

Figure 2.1: Dynamic viscosity measured for Hanks' balanced salt solution and compared to values of water and seawater and experimental data for lymphatic liquids.

In the experimental situation the biological liquids are usually replaced by Hanks' balanced salt solution (Sigma; supplemented with 4.1 mol·m^{-3} NaHCO$_3$ and 10 mol·m^{-3} HEPES buffer), with osmolarity between 300 and 320 mOsm (pH 7.35). The viscosity of this solution was measured with a rheometer (MCR 301, Anton Paar GmbH, Austria). The measurements were conducted by Loredana Pop (Technische Universität Dresden). The solution was taken in a composition as used in a guinea pig preparation (Nowotny and Gummer 2006). The experiments were conducted up to room temperature of around 22 °C. As no protection against evaporation was available, only measurements near room temperature were possible. The results are compared in Figure 2.1 with the literature values mentioned above and values for water (Kestin and Wakeham 1988) and seawater (Lide 2009) as well as the experimental data from Zwicker (1974) and from Dahl and Kleinfeldt (1970).

For a Newtonian fluid the viscosity is not a function of the deformation rate. To check whether this hypothesis can be applied to the Hanks' balanced salt solution, viscosity measurements with different shear rates were accomplished. This dependence is given in Figure 2.2. For lower shear rates the scatter of the measurement increases as the integral force decreases. In this range it is hard to identify trends. For high shear

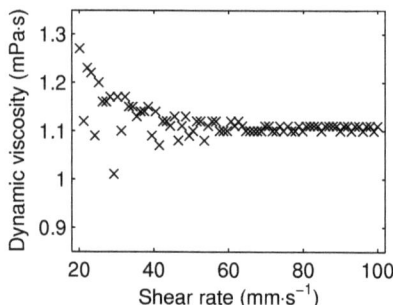

Figure 2.2: Dynamic viscosity measured for Hanks' balanced salt solution at 20 °C for different shear rates.

rates the viscosity is very constant. It is assumed that the Newtonian assumption of a shear stress being only proportional to the velocity is valid for inner-ear liquids.

The observation by Zwicker (1974) that a minimum pressure is necessary to drive the lymphatic liquid through a tube as well as the experimental observation of increasing viscosity with decreasing shear rate are indications of viscoelastic properties of the inner-ear liquids and need further verification.

To estimate the validity of a continuum approach the geometrical and molecular length are compared. The smallest reference length is taken as the wall-to-wall distance of two apposed stereocilia at the place of closest approach. In the model of this work this is ten nanometers. If the geometrical scale is smaller than lengths over which intermolecular forces are relevant, the continuum assumption for a liquid fails. The dimensions of the water molecule are in the order of one tenth of a nanometer. This fits about a hundred water molecules between two stereocilia at closest approach.

Another important geometrical measure is the Debye length, the distance over which charged-charged interactions play a role. This length

$$l_{\text{Debye}} = \sqrt{\frac{\varepsilon_0 \varepsilon_r k_B T}{2 N_A e^2 I}} \tag{2.37}$$

is based on the permittivity of free space ε_0, the Boltzmann constant k_B, the Avogadro number N_A, and the elementary charge e. For a typical ionic strength of $I = 300$ mol·m^{-3} for the inner-ear liquids, a room temperature of $T = 293.15$ K, and a dielectric constant of water $\varepsilon_r = 80$, the length over which ions interact is $l_{\text{Debye}} = 0.55$ nm which is more than ten times less the geometrical smallest size. The continuum assumption is supposed to be a valid approximation.

The bulk modulus of the lymph is so far not directly measured. As comparison values of water and seawater are given. All values are from the reference book by Lide

(2009). The seawater properties depend on the salinity. The values are taken for a value of $S = 35$, which is a typical concentration in the ocean. For a room temperature of 20 °C and a body temperature of about 40 °C the values are 2.2 GPa and 2.3 GPa for water. Seawater has a slightly higher bulk modulus. For 20 °C and 40 °C the values are 2.4 GPa and 2.5 GPa. These values are computed from the density and the speed of sound. The density variations are below 3 % for the temperature range of 20 °C to 40 °C for water with respect to a density value of 1 Mg·m^{-3}.

The thermal capacity of heat of water is 4.18 kJ·kg^{-1}·K^{-1} for the temperature range of 20 °C to 40 °C. For seawater this is 4.0 kJ·kg^{-1}·K^{-1}. For water the heat capacity ratio increases from 1.006 at 20 °C to 1.026 at 40 °C. The thermal conductivity increases for this temperatures from 0.60 W·m^{-1}·K^{-1} to 0.63 W·m^{-1}·K^{-1}. The coefficient of thermal expansion is given by the relation (2.35). For water at 20 °C it is 2.1×10^{-4} K^{-1} and 3.8×10^{-4} K^{-1} at 40 °C.

The water properties over a pressure range from 90 kPa to 110 kPa are inferior to the variations for the temperature range of 20 °C to 40 °C. The values are compared based on the values from Lemmon et al. (2010). The dependence on pressure of the properties is neglected.

If not stated otherwise, in this work representative and constant properties are used for the inner-ear liquids. These are for the dynamic viscosity $\eta = 1$ mPa·s, the density $\varrho = 1$ Mg·m^{-3} and for the bulk modulus $K = 1$ GPa. The Helmholtz number (2.48) for the hair bundle mechanics is always much smaller than one and thus the liquid is nearly incompressible. The bulk modulus is used to ensure the volume constancy and for simplicity just the order of magnitude is chosen.

2.2.3 Interaction of Liquid and Solid

The wall of the solid is impermeable for the liquid in normal direction. In the tangential direction the flow usually sticks fully to the wall. On scales of a few nanometers a partial slip can be observed depending on the properties of the liquid, those of the solid, and the geometry of the surface where they interact (Lauga et al. 2005). A way to account for the molecular effect at the surface of contact is by a slip length in continuum mechanical models. No slip-length measurements are known for the combination of stereocilia and endolymph; the precise geometry of the interstereociliary distances is also unknown. Due to lack of information possible slip at the wall is not considered here.

2.3 Simplified Equations for Inner-Ear Mechanics

Based on the above given material properties and physical scales, a dimensionless form of the equations of motion is used to evaluate the importance of the various terms in the differential equations. Order-of-magnitude estimates allow us to drop terms of minor importance.

2.3.1 Dimensionless Form of Fluid Equations

All variables are referenced by a characteristic value. The relations are

$$t^* = \omega t, \qquad \varrho^* = \frac{\varrho}{\varrho_{\text{ref}}},$$

$$\vec{v}^* = \frac{\vec{v}}{V}, \qquad T^* = \frac{T}{T_{\text{ref}}},$$

$$\nabla^* = l\, \nabla, \qquad \Phi^* = \frac{l^2}{\eta V^2}\, \Phi.$$

The reference values are indicated by the subscript $(\cdot)_{\text{ref}}$. The material properties are assumed to be constant. The reference of time is based on the reciprocal circular frequency $1/\omega$ of the oscillation.

The pressure is reduced by the average or ambient value p_{ref} and normalized by a representative pressure. In classical fluid mechanics this is usually the dynamic pressure ϱV^2. In the case of very low velocities this makes a minor contribution and the pressure results mainly from viscous stresses. They scale with the dynamic viscosity η divided by a time scale. Here for oscillatory flow with small amplitudes the dimensionless pressure

$$p^* = \frac{p - p_{\text{ref}}}{\eta\, \omega} \tag{2.38}$$

is referenced by dynamic viscosity and circular frequency. The subtraction of a reference pressure p_{ref} is without consequence for the dimensionless form as the pressure appears only in derivatives of space and time.

The equations to describe the fluid behavior are rewritten in non-dimensional form. It is assumed that the material properties are not dependent on position. Labels under the braces indicate the type of the corresponding terms in the equations. The expression for mass conservation (2.1) reads

$$\underbrace{\frac{\partial \varrho^*}{\partial t^*}}_{\text{inertial}} + \underbrace{\text{Kc}\, \vec{v}^* \cdot \nabla^* \varrho^*}_{\text{convective}} + \underbrace{\text{Kc}\, \varrho^* \nabla^* \cdot \vec{v}^*}_{\text{compressive}} = 0, \tag{2.39}$$

2.3 Simplified Equations for Inner-Ear Mechanics

momentum conservation (2.30) gives

$$\underbrace{\text{Wo}^2 \varrho^* \frac{\partial \vec{v}^*}{\partial t^*}}_{\text{inertial}} + \underbrace{\text{Re } \varrho^*(\vec{v}^* \cdot \nabla^*)\vec{v}^*}_{\text{convective}} =$$

$$\underbrace{-\frac{1}{\text{Kc}} \nabla^* p^*}_{\text{pressure}} + \underbrace{\nabla^* \cdot \left((\nabla^* \vec{v}^* + \vec{v}^* \nabla^*) - \frac{2}{3} \nabla^* \cdot \vec{v}^* \, \boldsymbol{I} \right)}_{\text{viscous–dissipative}}, \quad (2.40)$$

and energy conservation (2.31) implies

$$\underbrace{\text{Wo}^2 \varrho^* \frac{\partial T^*}{\partial t^*}}_{\text{inertial}} + \underbrace{\text{Re } \varrho^*(\vec{v}^* \cdot \nabla^* T^*)}_{\text{convective}} = \underbrace{\frac{1}{\text{Pr}} \nabla^* \cdot \nabla^* T^*}_{\text{thermal–dissipative}}$$

$$\underbrace{- \frac{\text{Ec Kt}}{\text{Kc}^2} T^* \frac{\partial p^*}{\partial t^*}}_{\text{compressive}} - \underbrace{\frac{\text{Ec Kt}}{\text{Kc}} T^* \vec{v}^* \cdot \nabla^* p^*}_{\text{convective–compressive}} + \underbrace{\text{Ec } \Phi^*}_{\text{viscous}} . \quad (2.41)$$

The dimensionless numbers Wo, Re, Kc, Pr, Ec, and Kt as well as the Helmholtz number He are discussed below.

Table 2.1: Order of magnitude estimates of physical quantities for hair bundle mechanics. The used units μnm are discussed in the Appendix A.

Quantity	Variable	Value
Length	l	10^0–10^2 μm
Circular frequency	ω	10^{-3}–10^0 ms^{-1}
Velocity	V	10^{-3} mm·s^{-1}
Dynamic viscosity	η	10^0 mPa·s
Density	ϱ_{ref}	10^{-3} Gg·m^{-3}
Temperature	T	10^2 K
Speed of sound	c	10^6 mm·s^{-1}
Thermal conductivity	k_{th}	10^9 nW·m^{-1}·K^{-1}
Heat capacity at constant pressure	c_p	10^9 μJ·kg^{-1}·K^{-1}
Coefficient of thermal expansion	β	10^{-4} K^{-1}

The estimation of the orders of magnitude is based on typical values listed in Table 2.1. The material properties of the endolymph were presented in Section 2.2.2. Additional parameters are needed to characterize the mechanics. For the geometry, one measure is the distance over which flow gradients develop and influence the liquid. Another is the diameter of single stereocilium, which is a little bit less than half a micrometer. A large geometrical scale is the distance between bundles which measures several tens of micrometers. For the lower time scale the observation is that the bullfrog sacculus is sensitive to oscillations starting from about ten hertz. Human hearing defines an upper limit of more than ten kilohertz. The objective of this work is to understand the hearing at threshold. Typical displacements for the bundle are in the order of a nanometer (Hudspeth 1989) at threshold for frequencies around one kilohertz. This is equivalent to velocities of the order of a micrometer per second.

These observations motivate the selected orders of magnitude. They are further supported by the experimental observations from Kozlov et al. (2007), which are extensively used in this work. The dimensionless quantities are evaluated for their minimal and maximal possible value with respect to the given ranges of the quantities.

Womersley number Womersley (1955) analyzed pulsating blood flow by decomposition into the spectral content of sinusoidal motions. The flow can then be characterized by the ratio of geometrical size, here the tube radius, to the length over which the momentum dissipates in a cycle

$$\text{Wo} = l\sqrt{\frac{\omega \varrho}{\eta}}. \tag{2.42}$$

For a sinusoidal pressure gradient in a circular tube the flow changes from a parabolic shape to plug flow for small to larger Womersley numbers. This number is also proportional to the square root of the ratio of inertia to viscous forces. Here this ratio reads

$$10^{-3} \lesssim \text{Wo} \lesssim 10. \tag{2.43}$$

Thus the inertial terms in the momentum and energy equations (2.40, 2.41) have to be retained as they might be of similar order as the viscous dissipation or thermal dissipation.

Reynolds Number The Reynolds number is the ratio of convective to dissipative momentum

$$\text{Re} = \frac{\varrho V}{\eta/l}. \tag{2.44}$$

For typical engineering problems this ratio is well above one. But a small geometrical scale and low velocity dramatically reduce this number. If this ratio is below one, the nonlinear convective terms in the momentum and energy equations disappear. This

2.3 Simplified Equations for Inner-Ear Mechanics

ratio is

$$10^{-6} \lesssim \mathrm{Re} \lesssim 10^{-4} \tag{2.45}$$

well below one. The convective terms on the left-hand side of the momentum and energy equations (2.40, 2.41) are therefore negligible. This greatly simplifies the system of equations as they are now linear in all variables.

Keulegan-Carpenter Number Keulegan and Carpenter (1958) analyzed the forces on cylinders and plates in oscillatory flow. To characterize the forces an important parameter is the ratio of two distances. The amplitude of the oscillatory motion of the fluid with respect to the body and the characteristic dimension of the body, which reads

$$\mathrm{Kc} = \frac{V/\omega}{l}. \tag{2.46}$$

If the velocity is large enough that parts of the fluid move completely around the body during one cycle, this ratio is larger than one. Here the fluid displacements are small compared to the geometry

$$10^{-5} \lesssim \mathrm{Kc} \lesssim 1. \tag{2.47}$$

If the frequency is low and the geometry small, this number is overestimated. At low frequencies the elastic force dominates and the mechanics are controlled by displacement rather than by velocity. Thus the effective velocity is smaller for low frequencies. It is assumed that the displacement amplitudes are always smaller than the geometrical dimensions and so $\mathrm{Kc} \ll 1$. Therefore, the convective and compressive terms are always smaller than the inertial term in the mass-conservation equation (2.39) and can be neglected. The pressure-gradient term in the momentum equation (2.40) is divided by this small number and thus is important. In the energy equation the Keulegan-Carpenter number is combined with other dimensionless numbers, which are discussed below.

Helmholtz Number The ratio of geometrical size to acoustic wavelength is the Helmholtz number

$$\mathrm{He} = \frac{l\,\omega}{c}. \tag{2.48}$$

For wavelengths shorter than the geometry, acoustic modes are present. For hair-bundle mechanics the Helmholtz number is well below one and therefore the fluid can be regarded as nearly incompressible

$$10^{-9} \lesssim \mathrm{He} \lesssim 10^{-4}. \tag{2.49}$$

Physical problems with acoustic resonances have numbers close to one.*

Number of Thermal Expansion The thermal expansion coefficient times the ambient temperature

$$\mathrm{Kt} = -\beta\, T \tag{2.50}$$

characterizes the variations of the volume strain with respect to temperature change. The value is

$$\mathrm{Kt} \approx 10^{-2}, \tag{2.51}$$

smaller than one. This implies minor volume change with respect to temperature variations. The number appears together with other dimensionless numbers which are discussed below.

Eckert Number The ratio of kinetic energy to the enthalpy is characterized by the Eckert number

$$\mathrm{Ec} = \frac{V^2}{c_p\, T}. \tag{2.52}$$

As a flow decelerates the temperature increases. Due to the very low velocities this ratio is

$$\mathrm{Ec} \approx 10^{-18}. \tag{2.53}$$

The last three terms of the right hand side of the energy equation (2.41) are multiplied by this very small number. The products with the number of thermal expansion and the Keulegan-Carpenter number are also much smaller than one. So it is safe to assume that unsteady pressure variations and their convective transport can be neglected as well as the thermal energy coming from viscous losses in the energy equation (2.41).

Prandtl Number The dissipation by momentum and thermal conductivity are compared by the Prandtl number. It is based on material properties in the form

$$\mathrm{Pr} = \frac{c_p\, \eta}{k_{\mathrm{th}}}. \tag{2.54}$$

For oscillatory flow this is also the ratio of the thermal and viscous boundary-layer heights. In air this ratio is close to one and for water it is

$$\mathrm{Pr} \approx 10, \tag{2.55}$$

---

*As an example in Section 3.2.2 the impedance of a tube is computed in the frequency range where the first eigenfrequencies are relevant.

2.3 Simplified Equations for Inner-Ear Mechanics

larger than one. This implies a thinner boundary layer for the temperature field than for the velocity field at the wall. A way to interpret this is by comparing the dissipative terms of the momentum and the energy equations (2.40, 2.41). The dissipative term of the energy equation is smaller than that of the momentum equation by the reciprocal of the Prandtl number. Thus, the ratio of viscous to thermal dissipation is proportional to the Prandtl number.

For classical fluid mechanical problems the Reynolds number and the Keulegan-Carpenter number are much larger than one. The Prandtl number is only material dependent. For water it is larger than one and for air about one. The Eckert number is relevant if the kinetic energy of the fluid is sufficient to heat the fluid; this usually plays a role in hypersonic flows. This motivates a simplification of the set of governing equations for special conditions such as the fluid mechanics in the inner ear.

2.3.2 Small-Amplitude Fluid Equations

Using the order of magnitude estimates allows to drop some terms in the equations describing the physics of the fluid (2.39, 2.40, 2.41). The mass conservation equation simplifies to

$$\frac{\partial \varrho}{\partial t} = 0 \tag{2.56}$$

as the Keulegan Carpenter number is small. The pressure change is related to the density change by the bulk modulus and density (2.13). The relation rewritten in dimensionless form is

$$\frac{\partial p^*}{\partial \varrho^*} = \frac{\mathrm{Wo}^2}{\mathrm{He}^2} . \tag{2.57}$$

This ratio is a function of the circular frequency and material properties as

$$\frac{\mathrm{Wo}^2}{\mathrm{He}^2} = \frac{K}{\omega \eta} . \tag{2.58}$$

For the given values yields

$$10^9 \lesssim K/(\omega \eta) \lesssim 10^{12} . \tag{2.59}$$

This ratio is much larger than one. Thus, the approximation that density variations over time are negligible does not imply that pressure variations can be neglected as well.

The momentum equation reads

$$\varrho \frac{\partial \vec{v}}{\partial t} = -\nabla p + \eta \nabla \cdot \left((\nabla \vec{v} + (\nabla \vec{v})^\mathsf{T}) - \frac{2}{3} \nabla \cdot \vec{v} \, I \right) \tag{2.60}$$

where the convective term proportional to the Reynolds number is omitted. The divergence of the velocity, the last term on the right-hand side, is in dimensionless form and integrated once over time with the pressure velocity relation of the fluid (2.24)

$$\nabla^* \cdot \vec{u}^* = -\frac{1}{K/(\omega\,\eta)}\, p^* \,. \tag{2.61}$$

This is always much smaller than one and so the divergence term of the velocity can also be neglected for the dissipation.

The energy equation simplifies to

$$\frac{\partial T}{\partial t} = 0 \tag{2.62}$$

as the thermal dissipation is smaller than viscous dissipation by one over the Prandtl number. The terms multiplied by the Eckert and Reynolds number are negligible. Thus the processes can be regarded as isothermal.

The simplified energy (2.62) and mass conservation (2.56) equations state that the temperature and density are constant over time. In the momentum equation (2.60) the variables of velocity and pressure appear. To complete the problem a pressure velocity relation is needed. The material law provides the missing link with the relation of volume strain and pressure (2.24) differentiated with respect to time

$$\frac{\partial p}{\partial t} = -K\,\nabla \cdot \vec{v} \,. \tag{2.63}$$

The equations above are derived under consideration of dimensionless numbers. Assuming small variations of the variables around a mean value leads to the same equations, as presented for example by Pierce (1994).[*]

The dimensionless analysis restricts the validity to situations where all the conditions

$$\text{Pr} \gg 1 \,, \quad \text{Re} \ll 1 \,, \quad \text{Ec} \ll 1 \,, \quad \text{Kc} \ll 1 \,, \text{ and } \quad \text{Ec}\,\text{Kt} \ll \text{Kc}^2$$

are satisfied.

2.3.3 Small-Amplitude Solid Equations

The solid material is purely elastic and without any dissipation (2.28) as discussed in Section 2.2.1. Assuming isothermal processes, which is also a valid assumption for the fluid as discussed above, the momentum equation suffices to solve the field problem

---

[*]For the velocity the mean value has to vanish, as otherwise the non-linear convective terms will not vanish.

2.3 Simplified Equations for Inner-Ear Mechanics

of the solid material. With the material law for the elastic isotropic material (2.28) the momentum equation (2.2) reads

$$\varrho \frac{\partial^2 \vec{u}}{\partial t^2} = \nabla \cdot \left(\mu_{\text{solid}} \left(\nabla \vec{u} + (\nabla \vec{u})^{\mathsf{T}}\right) + \lambda_{\text{solid}} \nabla \cdot \vec{u} \, \boldsymbol{I}\right) , \quad \vec{x} \in \Omega_{\text{solid}} \tag{2.64}$$

where the convective term of the substantial derivative is neglected. The displacements are small and also for the solid domain the Keulegan-Carpenter number is smaller than one.

2.3.4 Coupled Fluid and Solid Equations for Small Amplitudes

For the fluid the equations (2.60) and (2.63) with the variables velocity and pressure have to be solved. The solid is described by the displacements (2.64). The equations are solved in the domain Ω, which is subdivided into the fluid part Ω_{fluid} and the solid part Ω_{solid} (Figure 2.3). The boundary Γ of the domains is subdivided into the part between fluid and exterior Γ_{fluid}, solid and exterior Γ_{solid}, and fluid and solid Γ_{fsi}. At the outer boundary $\Gamma_{\text{fluid}} \cup \Gamma_{\text{solid}}$ boundary conditions have to be applied as displacements or forces. The boundary conditions complete the boundary value problem. This states a fluid-structure interaction problem.*

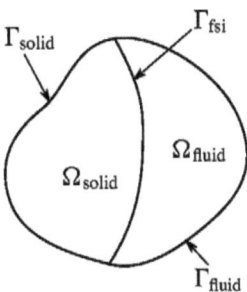

Figure 2.3: Regions of fluid and solid domain and their boundaries.

To solve the two-field problem a classical approach is to use a staggered algorithm where for each time step subiterations are necessary. In the subiterations the field problems of solid and fluid are solved independently and coupled by boundary conditions. The variables of velocity and pressure for the fluid and the displacement for the solid are retained. For the interface of fluid and solid Γ_{fsi} two continuities have to be satisfied.

---

*It would be more appropriate to call it fluid-solid interaction, but to be consistent with other publications the name fluid-structure interaction is used.

The velocity vectors have to be equal

$$\frac{\partial \vec{u}_{\text{solid}}(\vec{x})}{\partial t} = \vec{v}_{\text{fluid}}(\vec{x}) , \quad \vec{x} \in \Gamma_{\text{fsi}} \tag{2.65}$$

as well as the projection of the stress tensor on the normal direction of the interface surface $\vec{n}$

$$\boldsymbol{\sigma}(\vec{x})_{\text{solid}} \cdot \vec{n}_{\text{solid}} = \boldsymbol{\sigma}(\vec{x})_{\text{fluid}} \cdot \vec{n}_{\text{fluid}} , \quad \vec{x} \in \Gamma_{\text{fsi}} . \tag{2.66}$$

The positive normal direction is defined as positive pointing inwards to the domain Ω_{fluid} and Ω_{solid}.

One way to circumvent different variables for the fluid and solid is to describe the solid material in the context of fluid mechanics by the variable pressure and velocity (Greenshields and Weller 2005; Papadakis 2008).

For the small amplitudes encountered in hearing mechanics a different approach is pursued here. The velocity in the fluid equations is replaced by the time derivative of the displacement (2.7). The momentum balance for the fluid (2.60) reads

$$\boxed{\varrho \frac{\partial^2 \vec{u}}{\partial t^2} = -\nabla p + \eta \nabla \cdot \left(\left(\nabla \frac{\partial \vec{u}}{\partial t} + \left(\nabla \frac{\partial \vec{u}}{\partial t} \right)^{\mathrm{T}} \right) - \frac{2}{3} \nabla \cdot \frac{\partial \vec{u}}{\partial t} \boldsymbol{I} \right) , \quad \vec{x} \in \Omega_{\text{fluid}}} \tag{2.67}$$

and for the pressure (2.63)

$$\boxed{p = -K \nabla \cdot \vec{u} , \quad \vec{x} \in \Omega_{\text{fluid}}} . \tag{2.68}$$

This formulation in terms of displacement and pressure is convenient for coupling with elastic materials as it uses the displacement as common variable. Furthermore, the same fundamental principle of momentum conservation is used for the derivation.

To sum up, the fluid-structure interaction problem at small amplitudes is described by the same set of fundamental equations. For the solid the Equation (2.64) and for the fluid the Equation (2.67) have to be solved. Only the material law differs. Therefore the continuity conditions of displacement and normal stress are already incorporated into the differential equations. For the fluid an additional equation is solved for the pressure (2.68). This additional equation reduces possible issues with the numerical implementation as the large ratio $K/(\omega \eta)$ is not directly involved in the equations. As boundary condition the variables either displacements or forces are necessary to be defined at the outer boundary.

Chapter 3

Finite-Element Formulation for the Fluid

The finite-element method is employed to discretize the differential equations for the fluid. For the solid material a finite element formulation based on displacement is state of the art and not detailed here (Zienkiewicz et al. 2006). In this chapter a brief summary of the final matrix formulation is given. The implementation is validated by comparison to analytical solutions. The test cases were selected to be representative for the fluid mechanics in the inner ear and to satisfy the constraints of the dimensionless numbers presented in the previous chapter.

3.1 Finite-Element Discretization

The finite-element method is a standard technique in mechanics to solve field problems on complex geometries. The implementation is based on the *two-field incompressible elasticity (u-p form)* from the textbook of Zienkiewicz et al. (2006).[*] Three-dimensional, cartesian coordinates with the directions x, y, and z will be used here.

3.1.1 Galerkin Projection

The momentum equation of the fluid (2.67) weighted with a vectorial ansatz function $\vec{w}_u$ and integrated over the whole domain Ω

$$\int_\Omega \vec{w}_u \cdot \left(\varrho \ddot{\vec{u}} - \nabla \cdot \boldsymbol{\tau} + \nabla p \right) d\Omega = 0 \qquad (3.1)$$

---

[*]The pressure is regarded as positive for a decrease in volume in fluid mechanics. Zienkiewicz et al. (2006) use the definition of positive pressure for an increase of volume.

is the basis for a finite-element discretization. Herein the time derivatives are abbreviated by

$$\dot{\vec{u}} \equiv \frac{\partial \vec{u}}{\partial t} \quad \text{and} \quad \ddot{\vec{u}} \equiv \frac{\partial^2 \vec{u}}{\partial t^2} \,. \tag{3.2}$$

The deviatoric stress $\boldsymbol{\tau}$ is not rewritten in the displacements as in Equation (2.67).

To the integral expression the divergence theorem is applied

$$\int_\Omega \left(\vec{w}_u \cdot \varrho \ddot{\vec{u}} + \nabla \vec{w}_u : \boldsymbol{\tau} - \nabla \cdot \vec{w}_u \, p \right) d\Omega = \int_\Gamma \left(\vec{w}_u \cdot \vec{n} \cdot \boldsymbol{\tau} - p \, \vec{w}_u \cdot \vec{n} \right) d\Gamma \tag{3.3}$$

and higher order derivatives are thus shifted from the displacements to the ansatz function.* The normal vector $\vec{n}$ on the boundary Γ points into the domain Ω.

In addition the pressure displacement relation (2.68) is necessary to complete the problem. Weighted by the scalar ansatz function w_p this reads

$$\int_\Omega w_p \left(\nabla \cdot \vec{u} + \frac{1}{K} p \right) d\Omega = 0 \,. \tag{3.4}$$

To facilitate the algebraic implementation of the numerical solution of the field problem, the tensor notation is converted into a vector notation for the strains and stresses. Using Voigt notation the cartesian indices are replaced as

$$xx \to 1, \quad yy \to 2, \quad zz \to 3, \quad yz \to 4, \quad zx \to 5, \quad xy \to 6 \tag{3.5}$$

and thus the two dimensional symmetric strain tensor transforms into a vector

$$\boldsymbol{\varepsilon} \to \mathbf{e} \tag{3.6}$$

which has the components

$$\begin{pmatrix} \varepsilon_{xx} & \varepsilon_{xy} & \varepsilon_{zx} \\ \varepsilon_{xy} & \varepsilon_{yy} & \varepsilon_{yz} \\ \varepsilon_{zx} & \varepsilon_{yz} & \varepsilon_{zz} \end{pmatrix} \to \begin{bmatrix} e_1 \\ e_2 \\ e_3 \\ e_4 \\ e_5 \\ e_6 \end{bmatrix} . \tag{3.7}$$

The strain rates

$$\dot{\boldsymbol{\varepsilon}} \to \dot{\mathbf{e}} \tag{3.8}$$

---

*Pozrikidis (2005) provides some detail and practical explanations from the perspective of fluid mechanics.

3.1 Finite-Element Discretization

and deviatoric stresses

$$\tau \rightarrow \mathbf{T_d} \tag{3.9}$$

are similarly transformed. The strains are now computed from the displacement vector by the operator

$$\mathcal{S} = \begin{bmatrix} \partial_x & 0 & 0 \\ 0 & \partial_y & 0 \\ 0 & 0 & \partial_z \\ 0 & \partial_z & \partial_y \\ \partial_z & 0 & \partial_x \\ \partial_y & \partial_x & 0 \end{bmatrix} \tag{3.10}$$

as

$$\mathbf{e} = \mathcal{S}\vec{u} \: . \tag{3.11}$$

Based on the vector notation the relations of the material law are rewritten for the fluid. The pressure displacement relation (2.24) reads

$$p = -K\,\mathbf{m}^\mathsf{T}\,\mathbf{e} \tag{3.12}$$

with

$$\mathbf{m} = \begin{bmatrix} 1 \\ 1 \\ 1 \\ 0 \\ 0 \\ 0 \end{bmatrix} \tag{3.13}$$

to extract the trace from the strain **e**. For the deviatoric stresses τ of the fluid (2.23) the relation is rewritten as

$$\mathbf{T_d} = 2\,\eta\,\mathbf{I_d}\,\dot{\mathbf{e}} \tag{3.14}$$

with the deviatoric projection matrix

$$\mathbf{I_d} = \begin{bmatrix} 2/3 & -1/3 & -1/3 & 0 & 0 & 0 \\ -1/3 & 2/3 & -1/3 & 0 & 0 & 0 \\ -1/3 & -1/3 & 2/3 & 0 & 0 & 0 \\ 0 & 0 & 0 & 1/2 & 0 & 0 \\ 0 & 0 & 0 & 0 & 1/2 & 0 \\ 0 & 0 & 0 & 0 & 0 & 1/2 \end{bmatrix} \tag{3.15}$$

to extract the deviatoric part from the strain rates $\dot{e}$.

The relation for the deviatoric stresses (3.14) is used in the integral equation (3.3) and reads in vector notation

$$\varrho \int_\Omega \vec{w}_u^T \ddot{\vec{u}} \, d\Omega + 2\eta \int_\Omega \vec{w}_u^T \mathcal{S}^T \mathbf{I_d} \, \mathcal{S} \dot{\vec{u}} \, d\Omega - \int_\Omega \vec{w}_u^T \mathcal{S}^T \mathbf{m} \, p \, d\Omega = \int_\Gamma \vec{w}_u^T \vec{f} \, d\Gamma \quad (3.16)$$

where in $\vec{f}$ the forces on the boundary Γ are summarized. This is the complete right-hand side of Equation (3.3). The material properties are regarded as constant over the domain Ω.

For the pressure (3.4) the integral equation with the expression for the bulk modulus (3.12) reads

$$-\int_\Omega w_p \mathbf{m}^T \mathcal{S} \vec{u} \, d\Omega - \frac{1}{K} \int_\Omega w_p \, p \, d\Omega = 0 \, . \quad (3.17)$$

This equation has no term with a boundary integral.

Displacement and pressure are approximated by

$$\vec{u} \approx \mathbf{N_u} \mathbf{u} \quad \text{and} \quad p \approx \mathbf{N_p} \mathbf{p} \, , \quad (3.18)$$

basis functions with local support for the entire domain Ω. The degrees of freedom are gathered in $\mathbf{u}$ for the displacement and in $\mathbf{p}$ for the pressure.

The velocities and accelerations are approximated by the same basis functions as the displacements as:

$$\dot{\vec{u}} \approx \mathbf{N_u} \dot{\mathbf{u}} \quad \text{and} \quad \ddot{\vec{u}} \approx \mathbf{N_u} \ddot{\mathbf{u}} \, . \quad (3.19)$$

If the ansatz function is replaced by the basis function of the displacement, a symmetric formulation is possible. The Equation (3.16) than reads in vector notation

$$\varrho \int_\Omega \mathbf{u}^T \mathbf{N_u}^T \mathbf{N_u} \ddot{\mathbf{u}} \, d\Omega + 2\eta \int_\Omega \mathbf{u}^T \mathbf{B_u}^T \mathbf{I_d} \mathbf{B_u} \dot{\mathbf{u}} \, d\Omega$$
$$- \int_\Omega \mathbf{u}^T \mathbf{B_u}^T \mathbf{m} \mathbf{N_p} \mathbf{p} \, d\Omega = \int_{\Gamma_t} \mathbf{u}^T \mathbf{N_u}^T \mathbf{g} \, d\Gamma \quad (3.20)$$

with $\mathbf{B_u}$ introduced as

$$\mathbf{B_u} = \mathcal{S} \mathbf{N_u} \, . \quad (3.21)$$

The pressure displacement relation (3.17) reads equivalently

$$-\int_\Omega \mathbf{p}^T \mathbf{N_p}^T \mathbf{m}^T \mathbf{B_u} \mathbf{u} \, d\Omega - \frac{1}{K} \int_\Omega \mathbf{p}^T \mathbf{N_p}^T \mathbf{N_p} \mathbf{p} \, d\Omega = 0 \quad (3.22)$$

3.1 Finite-Element Discretization

where the ansatz function is set to the basis function of the pressure to maintain symmetry.

This system of equations for the fluid is combined in a single matrix equation. Furthermore, the constant material properties incorporate the inertial force proportional to the density ϱ, the shear stresses scaled by the dynamic viscosity η, and the pressure being equivalent to the negative volume strain by the bulk modulus K.

A valid ansatz function with a single degree of freedom differs from zero. One possible choice is to take the unity matrix to represent all possible combinations. This leads to the following system of equations

$$\begin{bmatrix} \mathbf{M}_{uu} & \mathbf{0}_{up} \\ \mathbf{0}_{pu} & \mathbf{0}_{pp} \end{bmatrix} \begin{bmatrix} \ddot{\mathbf{u}}_u \\ \ddot{\mathbf{p}}_p \end{bmatrix} + \begin{bmatrix} \mathbf{C}_{uu} & \mathbf{0}_{up} \\ \mathbf{0}_{pu} & \mathbf{0}_{pp} \end{bmatrix} \begin{bmatrix} \dot{\mathbf{u}}_u \\ \dot{\mathbf{p}}_p \end{bmatrix} + \begin{bmatrix} \mathbf{0}_{uu} & \mathbf{K}_{up} \\ \mathbf{K}_{pu} & \mathbf{K}_{pp} \end{bmatrix} \begin{bmatrix} \mathbf{u}_u \\ \mathbf{p}_p \end{bmatrix} = \begin{bmatrix} \mathbf{f}_u \\ \mathbf{0}_p \end{bmatrix} \quad (3.23)$$

where $\mathbf{0}$ denotes a zero matrix and the subscripts $(\cdot)_u$ denote the degrees of freedom related to the displacement, velocity, and acceleration and $(\cdot)_p$ those of the pressure. Herein are the mass matrix for the accelerations

$$\mathbf{M}_{uu} = \varrho \int_\Omega \mathbf{N}_u^T \mathbf{N}_u \, d\Omega \, , \quad (3.24)$$

the damping matrix for the velocities

$$\mathbf{C}_{uu} = 2\eta \int_\Omega \mathbf{B}_u^T \mathbf{I}_d \mathbf{B}_u \, d\Omega \, , \quad (3.25)$$

the stiffness matrix for the pressure

$$\mathbf{K}_{pp} = -\frac{1}{K} \int_\Omega \mathbf{N}_p^T \mathbf{N}_p \, d\Omega \, , \quad (3.26)$$

and the matrix coupling pressure and displacement

$$\mathbf{K}_{up} = \mathbf{K}_{pu}^T = -\int_\Omega \mathbf{B}_u^T \mathbf{m} \mathbf{N}_p \, d\Omega \, . \quad (3.27)$$

The shape functions for the displacement $\mathbf{N}_u$ are quadratic without nodes in the center of the faces and in the center of the element (Figure 3.1).[*] The shape functions for the pressure $\mathbf{N}_p$ are linear. The displacement and pressure shape functions are both continuous from element to element.[†] This difference in degrees of freedom is necessary to satisfy the Babuška-Brezzi condition. The geometry of the finite element uses the same shape functions as the displacements to map the unit element coordinates onto the global geometry. The integration is performed by the complete Gaussian quadra-

---

[*]This type of element is a member of the serendipity family.
[†]Zienkiewicz et al. (2006) call this type of element in two dimensions Q 8/C4.

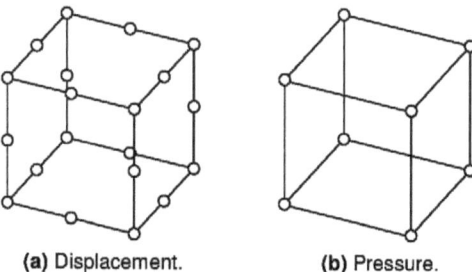

(a) Displacement.   (b) Pressure.

Figure 3.1: Finite element with node locations of the shape functions.

ture on the unit element. Throughout this work meshes consisting only of hexahedrons are used.

3.1.2 Frequency Formulation

In hearing it is essential to look at periodic signals with an assigned frequency. It is common practice to use Fourier transform for the data analysis. For the numerical model an integration over a sinusoidal time signal is inefficient in comparison to a single solution for one frequency. As the system of equations is linear they can be directly Fourier transformed. The basis is an ansatz for the pressure as

$$p(\vec{x},t) = \mathrm{Re}\left(\hat{p}(\vec{x})\,e^{-i\omega t}\right) \tag{3.28}$$

and for the displacement as

$$\vec{u}(\vec{x},t) = \mathrm{Re}\left(\hat{\vec{u}}(\vec{x})\,e^{-i\omega t}\right) \tag{3.29}$$

where the circular frequency is $\omega = 2\pi f$ and i is the imaginary unit. The velocity is the time derivative of the displacement and consequently

$$\dot{\vec{u}}(\vec{x},t) = -\omega\,\mathrm{Re}\left(i\hat{\vec{u}}(\vec{x})\,e^{-i\omega t}\right)\;; \tag{3.30}$$

the acceleration is the second time derivative

$$\ddot{\vec{u}}(\vec{x},t) = -\omega^2\,\mathrm{Re}\left(\hat{\vec{u}}(\vec{x})\,e^{-i\omega t}\right)\;. \tag{3.31}$$

For convenience the hat $(\hat{\cdot})$ as well as the real-part operator $\mathrm{Re}\,(\cdot)$ will be omitted in subsequent writing when functions of frequency and time are not present in the same equation.

The matrix formulation of the finite-element discretization in the frequency domain reads

$$\left(-\omega^2 \begin{bmatrix} \mathbf{M}_{uu} & \mathbf{0}_{up} \\ \mathbf{0}_{pu} & \mathbf{0}_{pp} \end{bmatrix} - i\omega \begin{bmatrix} \mathbf{C}_{uu} & \mathbf{0}_{up} \\ \mathbf{0}_{pu} & \mathbf{0}_{pp} \end{bmatrix} + \begin{bmatrix} \mathbf{0}_{uu} & \mathbf{K}_{up} \\ \mathbf{K}_{pu} & \mathbf{K}_{pp} \end{bmatrix} \right) \begin{pmatrix} \mathbf{u}_u \\ \mathbf{p}_p \end{pmatrix} = \begin{pmatrix} \mathbf{f}_u \\ \mathbf{0}_p \end{pmatrix}. \qquad (3.32)$$

The Fourier transform is a powerful tool but has a drawback. The transformation is for periodic signals of indefinite duration, but in reality the stimulus is present for only a finite time. Therefore, it would be more appropriate to use a transform with finite time window, in example the wavelet transform. Although the signals are finite in hearing mechanics, they usually oscillate for hundreds of cycles and thus the Fourier transform is still an adequate tool for the analysis of inner ear mechanics.

3.1.3 Implementation

The finite-element code was successfully checked by a multiple-element patch test under Matlab (Version 7.7, The MathWorks Inc., Natick, Massachusetts). For larger problems this Matlab code was implemented by Mario Fleischer (Technische Universität Dresden) as user-defined element into the finite-element software Ansys (Version 11, Ansys Inc., Canonsburg, Pennsylvania). For validation purposes three problems related to hearing mechanics with an analytical solution are presented. These examples were also used to select a suitable mesh size to resolve the pressure and displacement field with sufficient accuracy.

3.2 Validation of the Finite-Element Formulation

3.2.1 Oscillating Sphere

Denk et al. (1989) use the drag coefficient of a sphere in water with a radius of 2.5 μm as reference to compare with a hair bundle. The same geometry is used to validate the correct implementation of the fluid element for three-dimensional oscillatory flow problems with viscous and inertial forces.

The analytical solution for an oscillating sphere by Stokes (1856) provides a valuable reference. As the method here, this analytical solution relies on the assumption of small displacements and oscillatory motion.

If a sphere is accelerated slowly, the surrounding fluid moves relatively and the liquid shear causes frictional forces. If the acceleration is faster, inertial forces intervene. The reaction force divided by the velocity at the same location in the same direction is the drag coefficient. For an investigation in the frequency domain the drag coefficient c_{sph} is a function of the circular frequency ω as

$$c_{sph} = m\,\omega\,k_{cof}. \qquad (3.33)$$

Herein the mass of the sphere with radius r and density ϱ reads

$$m = \frac{4}{3}\pi \varrho\, r^3 \tag{3.34}$$

and the coefficient

$$k_{\text{cof}} = \frac{1}{2} + \frac{9}{4\sqrt{2}}\left(\frac{1}{\text{Wo}} - i\left(\frac{1}{\text{Wo}} + \frac{1}{\sqrt{2}\,\text{Wo}^2}\right)\right) \tag{3.35}$$

where i is the imaginary unit. The Womersley number (2.42) is based on the radius of the sphere and also a function of the frequency.

This analytical solution is valid for an outer boundary of the liquid at infinity. With the standard finite-element method this cannot be meshed.* To analyze the influence of the outer boundary distance on the computed drag, a series of computations were performed with fluid-layer thickness ranging from the sphere radius of $r = 2.5$ μm to 100 times of the radius. The frequencies range from 1 Hz to 1 MHz. The properties for the liquid are similar to water with a bulk modulus of $K = 1$ GPa, density of $\varrho = 1$ Mg·m^{-3}, and a dynamic viscosity of $\eta = 1$ mPa·s. The sphere around which the fluid moves has a zero density and a perfectly rigid wall. The analytical solution for the steady drag coefficient, which is equivalent to the limit of the frequency towards zero, is $c_{\text{sph}} = 6\pi\eta r$ and with the given values $c_{\text{sph}} = 47$ nN·s·m^{-1}.

On the artificial outer boundary of the liquid, two boundary conditions were applied. The first case presented in Figure 3.2 is with zero forces applied at the outer boundary. In the second case, shown in Figure 3.3, the outer spherical surface is blocked in motion for all directions. Both are extreme situations. The first one is equivalent to a change of density to zero at the outer boundary radius and the second to a sudden increase to infinite density.

If the outer boundary is free to move, the drag curve faithfully follows the reference solution down to the frequency where the unsteady boundary layer is larger than the fluid layer's thickness (Figure 3.2). If the frequency is lower, the relative motion declines and so does the drag. What remains is the inertial force needed to move the mass of the fluid. Consequently the phase lags by $\pi/2$ for low frequencies.

In the other case where the outer boundary is held in position, the inertial forces decrease with decreasing frequency (Figure 3.3). If the fluid thickness is not much larger than the sphere radius, the drag forces are higher due to the smaller distance over which shearing occurs. For high frequencies where the boundary layer is thinner than the fluid layer thickness, the numerical results align with the analytical solution.

To sum up, the numerical finite-element solution of an oscillating sphere in a viscous fluid converges to the analytical solution. The finite size of the mesh has to be large

---

*There are ways to circumvent this difficulty such as infinite elements, perfectly-matching layers, and multipole expansions.

3.2 Validation of the Finite-Element Formulation

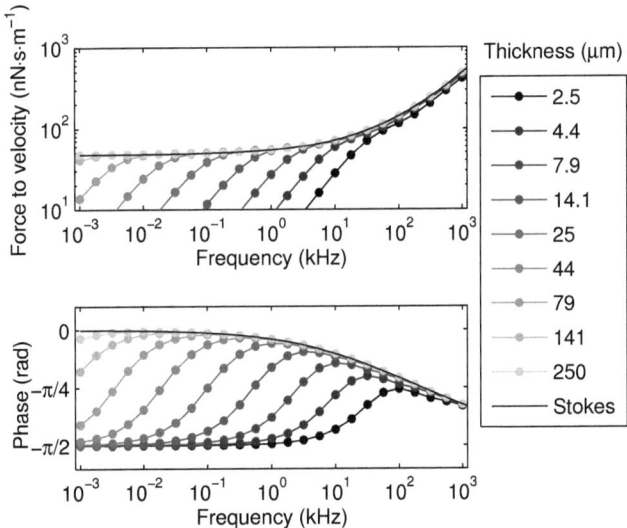

Figure 3.2: Drag coefficient of an oscillating sphere surrounded by a fluid layer with constant thickness. The thickness is given in the legend. The upper plot shows the absolute value and the lower plot the phase angle. No boundary conditions are applied to the outer spherical surface. The analytical reference solution is the black solid line from Stokes (1856).

enough to encompass the boundary layer. If the outer boundary is fixed, it is difficult to see from the drag coefficient an insufficient domain size. For an outer boundary which is free to move, a deficient fluid thickness appears as phase change.

3.2.2 Acoustic Tube

The method implemented to model the fluid is valid for viscous-acoustic problems with small amplitudes. This example presents a validation relevant for problems dominated by acoustic modes. The resonances in a tube are damped by the local gradient of the velocity at the wall. The problem was originally analyzed by von Helmholtz (1863). Later Kirchhoff (1868) investigated also the thermal losses that play a role in air. Because the energy equation is dropped, only the viscous losses are modeled here.

The properties used for the fluid are similar to those of water and the same as for the computation of the drag of a sphere (Section 3.2.1). The geometrical dimensions are close to the geometry of the cochlear chambers in the guinea pig (Thorne et al. 1999). The length of the tube is $l = 20$ mm and the radius $r = 0.4$ mm.

At the one end the tube is closed. On the other end a piston-like motion sinusoidally excites the fluid. If the excitatory frequency coincides with the resonance frequency,

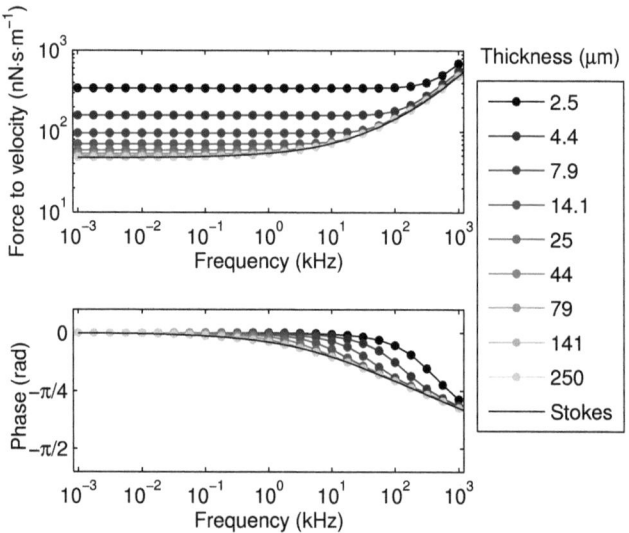

Figure 3.3: Drag coefficient of an oscillating sphere surrounded by a fluid layer with constant thickness. The thickness is given in the legend. The upper plot shows the absolute value and the lower plot the phase angle. The outer spherical surface is held in position. The analytical reference solution is the black solid line from Stokes (1856).

the pressure amplitude peaks at the inlet. To quantify this the impedance at the inlet is computed. This is the ratio of pressure to volume flow rate and is plotted in Figure 3.4

To avoid a mismatch of the velocities locally at the inlet, the analytical solution of the oscillating flow in a tube was applied as inlet velocity profile. The geometry was meshed with 50 elements over the length. To ensure a good resolution of the viscous boundary layer, the mesh was refined normally to the tube wall with a height ratio of 1.3 for successive layers of elements. The height measured normal to the wall of the first element at the boundary is

$$h_{bl} \approx 5 \frac{r}{\text{Wo}} \tag{3.36}$$

with the Womersley number based on the tube radius r. For every computed frequency step the mesh was newly generated to correctly resolve the unsteady boundary layer. The remaining interior of the tube was meshed with approximately equilateral elements. Only a quarter circle with symmetry constrains in circumferential direction was computed. The analytical solution given by Nederveen (1998) serves as reference. Without thermal losses the impedance of a closed tube is

$$Z = -\frac{i \varrho_w c_w}{\pi r^2} \cot(k_w l) \tag{3.37}$$

3.2 Validation of the Finite-Element Formulation

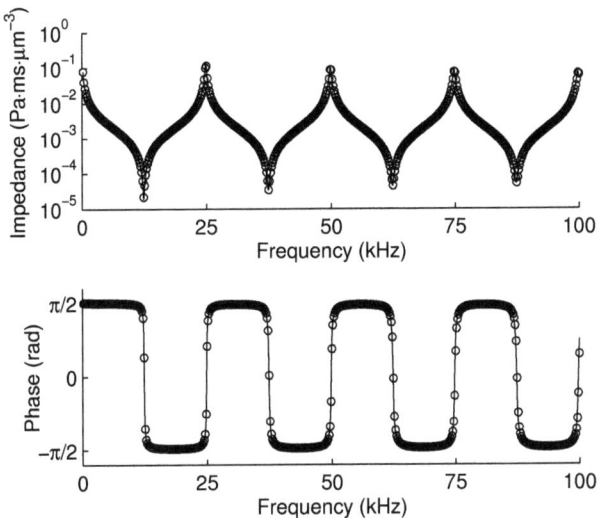

Figure 3.4: Impedance at the inlet of a tube closed at the end and filled with water. The tube length is 20 mm and the radius 0.4 mm. The upper plot shows the absolute value and the lower plot the phase angle. The markers are the numerical results and the solid line the analytical solution.

with the properties corrected for boundary-layer effects marked by the subscript $(\cdot)_w$. The modified density reads

$$\varrho_w = \varrho \left(1 + (1-i)\frac{\sqrt{2}}{Wo}\right) , \qquad (3.38)$$

the corrected wave number is

$$k_w = k\left(1 + \frac{(1-i)}{\sqrt{2}\,Wo}\right) \quad \text{with} \quad k = \frac{\omega}{c} \qquad (3.39)$$

and the speed of sound is modified to

$$c_w = \sqrt{\frac{K}{\varrho_w}} . \qquad (3.40)$$

The Womersley number is much larger than one with $Wo \approx 220$ for a frequency of 50 kHz. Therefore, the resonance peaks are sharp and the phase jumps over small frequency ranges.

Overall the method proves to be able to correctly compute acoustic problems with viscous dissipation. For fluids where the Prandtl number (2.54) is equal or below one, as for example air, it would be necessary to include the energy equation for thermal dissipation.

3.2.3 Fluid-Filled Elastic Sphere

The finite-element discretization should satisfy the continuity conditions for displacement (2.65) and stress (2.66) at the interface between solid and fluid. A spherical shell filled with fluid is used to validate the implementation. Motivated by the basal end of an outer hair cell, a sphere with an outer radius of 5 μm and a wall thickness of 0.1 μm is modeled. The interior of the spherical shell is filled with a water-like fluid. The properties used for the fluid are similar to water and the same as for the computation of the drag of a sphere (Section 3.2.1). The elastic shell is an isotropic material with a Young's modulus of $E_{\text{solid}} = 100$ kPa. The value is similar to the properties used by Tolomeo and Steele (1995) for an outer hair cell wall. The Poisson's ratio is set such that the bulk modulus of the solid is a scalar factor α times the bulk modulus of the fluid. This reads as

$$\nu(\alpha) = \frac{1}{2} - \frac{1}{6}\frac{E_{\text{solid}}}{K_{\text{fluid}}}\frac{1}{\alpha}. \tag{3.41}$$

The volumetric strain is applied by a constant displacement in radial direction on the outer boundary of the elastic sphere.

The mesh has 16 elements for a quarter circle. The elastic shell is resolved with two elements over the height. For the elastic material the hexahedral 20-node element Solid186 of Ansys (Version 11, Ansys Inc., Canonsburg, Pennsylvania) is used.[*] The mesh of the fluid is refined towards the elastic shell. The three dimensional model is just meshed as an eighth sphere with symmetry boundary conditions in the circumferential direction. The results are reported for a frequency of $f = 1$ Hz. At this low frequency the physical behavior is quasi-static.

In Figure 3.5a the negative pressure in the fluid and the radial stress of the elastic shell are plotted normalized by a reference pressure. This reference pressure is the bulk modulus of the fluid times the volume strain of the whole sphere. The volume strain of the sphere is the ratio of displacement at the outer boundary times the surface divided by the whole sphere volume. For equal bulk moduli in fluid and solid ($\alpha = 1$) the volume strains are as well close to each other.[†] If the outer shell is softer, the pressure is dominated by the bulk modulus of the elastic material.

In Figure 3.5b the displacement in the radial direction is plotted for various values of α normalized by the amplitude at the outer boundary u_{ref}. If the ratio α is one, the radial displacement is a linear function going through the origin. For smaller values

---
[*]For an accurate solution it was necessary to use the full integration option.
[†]In the solid material the radial and circumferential stress are not equal for Poisson's ratios smaller than 1/2. For the material parameters used, they are close to each other: $\nu \approx 0.499983$.

3.2 Validation of the Finite-Element Formulation

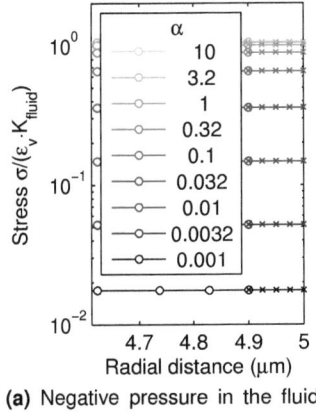

(a) Negative pressure in the fluid and radial stress in the solid.

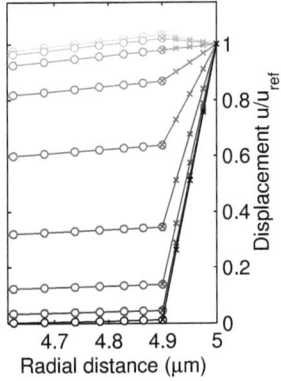

(b) Displacement in radial direction.

Figure 3.5: Elastic shell filled with a fluid displaced at the outer boundary in radial direction. The results of the fluid are marked by circles (o) and of the solid by crosses (×). The ratio of the bulk moduli is $\alpha = K_{\text{solid}}/K_{\text{fluid}}$ and given in the legend. At the radial distance of 4.9 μm the stresses and displacements are continuous.

the compression takes place mainly in the softer elastic shell and the fluid tends to zero displacement. For higher values of α the outer elastic shell becomes incompressible. The surface decreases with decreasing radius and so a nearly incompressible material has higher amplitudes of motion towards the center. The area changes from the outer to the inner radius of the elastic sphere by about 5 %.

Hereby the method proofed to be successful also for the interaction of solid and fluid, albeit without shear deformation. The normal stresses are correctly transferred between solid and fluid as well as the displacements. In the test case with the oscillating sphere shear stresses contribute significantly to the reaction force at the boundary and were correctly computed. Thus, the fluid reaction forces are correctly treated and interaction forces between solid and fluid are correctly transferred. Henceforth a validated numerical method is ready for computation of fluid-structure interaction problems in the inner ear.

Chapter 4

Model of the Hair Bundle from the Bullfrog's Sacculus

Kozlov et al. (2007) measured with two laser interferometers simultaneously the motion at the tall and short edges of the bullfrog's sacculus hair bundle in a thermal bath. The experimental data show a high coherence for frequencies from a few hertz up to several kilohertz. This suggests that the mechanical coupling between the stereocilia is stronger than the resistance of individual stereocilia in the bundle against deflection. The main objective here is to investigate to what degree the viscous liquid between the stereocilia mechanically couples them and what amount of elastic coupling might be needed to reproduce the experimental observations.

In this chapter the geometry of the hair bundle from the bullfrog's sacculus is used to build a detailed numerical model. This model encompasses the mechanics of the elastic properties of the cellular structure and the viscous liquid surrounding the bundle and filling the gaps between the stereocilia. The mechanical parameters of the model are adjusted to match different experimental observations.

The numerical model of the hair bundle relies on the equations describing the relevant physics, the geometry, and the material properties. The equations were discussed in Chapter 2 and a numerical method to solve them on complex geometries was presented in Chapter 3. Literature and microscopic images provide the necessary geometrical information for the bundle model. Experimental data are the reference for the mechanical response and to estimate and adjust the unknown material properties.

4.1 Geometry

The careful morphologic study of the bullfrog's sacculus by Jacobs and Hudspeth (1990) provides most of the required information. The geometry for the model is idealized and adjusted to simplify the geometry setting and the meshing procedure. The dimensions are taken as close as possible to the experimental observations.

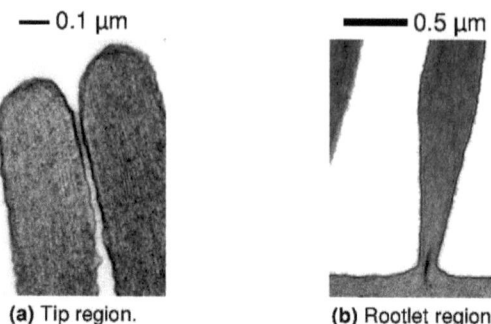

Figure 4.1: Transmission electron micrograph of different regions of stereocilia. In the tip region is the top connector material visible as darker area in the gap between the stereocilia. *Courtesy of Dr. A. J. Hudspeth, Howard Hughes Medical Institute*

The model was constructed with the stereocilia placed on a hexagon with $n_{\text{edg}} = 5$ stereocilia on each side. For the full hexagon the total number of stereocilia is $n_{\text{sc}} = 3\, n_{\text{edg}}\, (n_{\text{edg}} - 1) + 1 = 61$. At the tall edge the stereocilium is replaced by the kinocilium and the three next neighbors are removed. Thus, the total number of stereocilia is 57 and there is one kinocilium. Figure 4.2a shows this arrangement in the top view. By this five stereocilia are the next neighbors to the kinocilium, as experimentally observed.* The longest stereocilium has a length of about 7.8 μm. For simplicity the height of the kinocilium measures $h_{\text{kc}} = 8.45$ μm to keep the height profile smooth over the whole bundle. The kinociliary bulb is placed at the height of $h_{\text{kb}} = 6.5$ μm. The height decreases from one stereocilium to the next shorter by $h_{\text{t}} = 325$ nm. The diameter of the stereocilia and the kinocilium is $d_{\text{sc}} = 380$ nm. At their tips all are rounded by a half sphere with a radius equal to that of the cylindrical part.†

At the bottom section the diameter decreases towards the base to $d_{\text{b}} = 125$ nm. The height of the rootlet section, the conical frustrum, is set to $h_{\text{r}} = 3\, h_{\text{t}} = 975$ nm. These data were estimated from micrographs as given in Figure 4.1b.

The stereocilia and the kinocilium are interconnected by different links, as sketched in Figure 4.2. The kinocilium is connected to the next neighboring stereocilia by kinociliary links at about the height of the kinociliary bulb. There are no kinociliary links between stereocilia. They are interconnected at their top by single, oblique tip

---

*The enlarged distance between the kinocilium and the stereocilia in the model was helpful. The bending kinocilium is clamped at the point of insertion and displaces with a quadratic displacement profile over the height. The neighboring pivotal stereocilia are just bending in the rootlet section next to the point of insertion, above this bending section they displace linearly versus the height. By the kinociliary links their displacements are coupled next to the tips. The increased distance between stereocilia and kinocilium ensures that the linear and quadratic displacement profiles do not collide next to the rootlet section.

†García et al. (1998) observe an increasing diameter of the stereocilia from the tip to the bottom. This possible diameter variation is not taken into account as it was not observed in all micrographs of this study.

4.1 Geometry

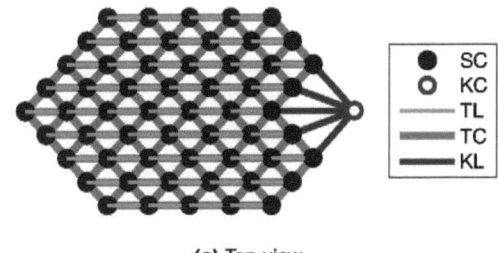

(a) Top view.

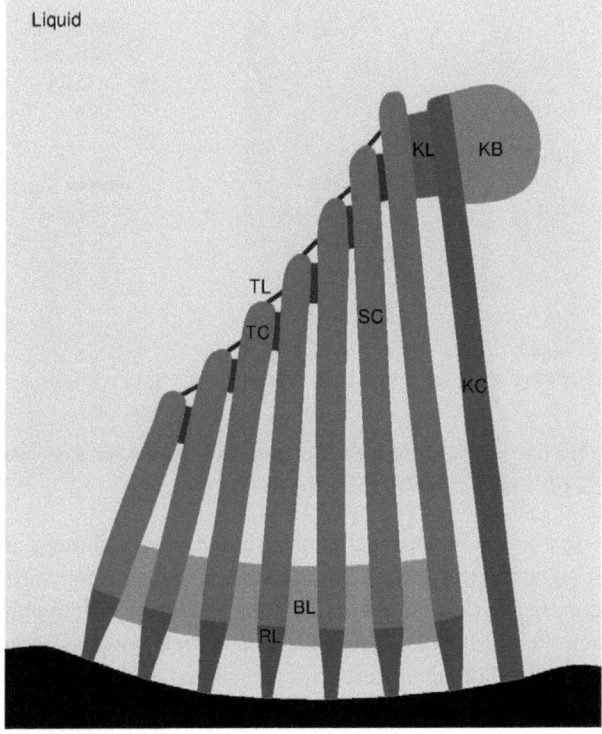

(b) Symmetry plane.

Figure 4.2: Schematic drawing of the hair bundle from the bullfrog's sacculus with the excitatory direction to the right. BL: Basal links, CP: Cuticular plate, KB: Kinociliary bulb, KC: Kinocilium, KL: Kinociliary links, RL: Rootlets, SC: Stereocilia, TC: Top connectors, TL: Tip links.

links. They link stereocilia only in the physiologically mechano-sensitive direction, the direction along which the bundle is sensitive to the stimulus. The angle between the tip-link and stereociliary axis is about 45°. All other links interconnect a stereocilium to all next neighbors.

The top connectors are just below the tip links in the narrow gap between the stereocilia. They spread over a height of about 0.5 µm (Figure 4.1a). Basal links connect the basal shafts of stereocilia.

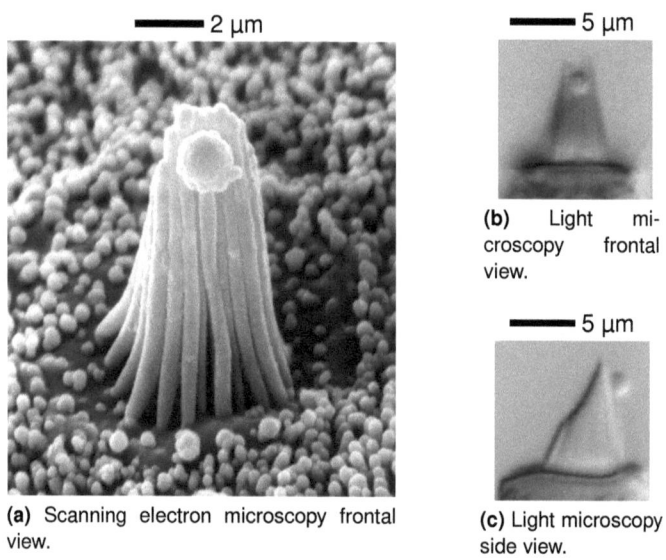

(a) Scanning electron microscopy frontal view.

(b) Light microscopy frontal view.

(c) Light microscopy side view.

Figure 4.3: The hair bundle of the bullfrog's sacculus in frontal and side view. *Courtesy of Dr. A. J. Hudspeth, Howard Hughes Medical Institute*

For a consistent three-dimensional geometrical model, additional information was taken from micrographs. Recordings based on electron microscopy are sharp and have a high resolution, but during the preparation the specimen is dehydrated and fixated. Figure 4.3a shows a hair bundle seen in the frontal view with a scanning electron microscope. Therein the individual stereocilia are curved and not straight as in the light micrographs (Figures 4.3b and 4.3c). The resolution of the light microscope is not sufficient to clearly resolve the individual stereocilia in the upper part of the bundle, but the liquid medium ensures a native environment keeping the cells alive and provides geometrical information without shrinkage. As far as possible geometrical data from light micrographs are used.

The kinematics of the hair bundle are dominated by the ratio of tip-link extension to hair-bundle deflection (Jacobs and Hudspeth 1990), which is in a first-order approximation characterized by the stereocilia separation at their bottom divided by

4.1 Geometry

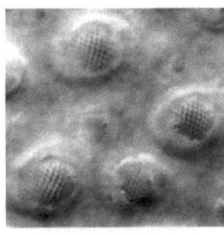

━━ 10 µm

Figure 4.4: Differential-interference-contrast micrograph of several hair bundles of a bullfrog's sacculus seen from top and kinocilium to the right. *Courtesy of Dr. A. J. Hudspeth, Howard Hughes Medical Institute*

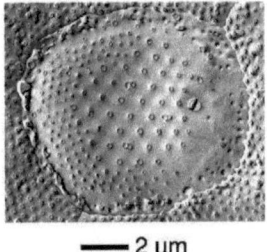

━━ 2 µm

Figure 4.5: Scanning electron micrograph of the cuticular plate with the stereocilia and kinocilium removed in a freeze-fractured preparation. The kinocilium is to the right. *Courtesy of Dr. A. J. Hudspeth, Howard Hughes Medical Institute*

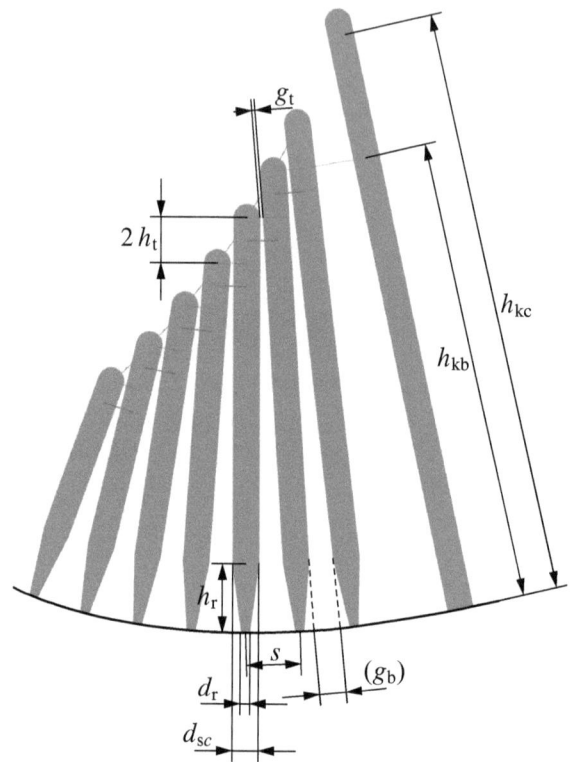

Figure 4.6: View of the symmetry plane of the bundle model with geometry parameters used in Table 4.1.

their height. The cuticular plate and stereocilia tips are clearly visible in the light microscopy side view and from this the angles and heights can be measured without artifacts. The measured center-to-center distance of the stereocilia on the cuticular plate might be erroneous by optical distortion or preparation treatment. Dr. A. J. Hudspeth (Howard Hughes Medical Institute) measured for the center-to-center spacing of the stereocilia on the cuticular plate with differential interference contrast microscopy $s_{dic} = 860 \pm 80$ nm ($n = 14$) from micrographs as in Figure 4.4. Care was taken in checking the absence of different following artifacts: The results are independent of the presence of the otolithic membrane. Also the refraction index of the liquid did not affect the results. The addition of 10 % metrizamide lowered the mismatch in index of refraction. The use of two different lenses (Achromat water-immersion objective lens, 63×, $NA = 0.9$ and C-Apochromat water-immersion objective lens, 63×, $NA = 1.2$) did not alter the results as well. The diameter of the stereocilia was reconstructed based

4.1 Geometry

Table 4.1: Geometrical parameters of the model of the hair bundle from the bullfrog's sacculus.

Name	Variable	Value	Ref.
Height of kinocilium	h_{kc}	8.45 μm	*
Height of kinociliary bulb	h_{kb}	6.5 μm	*
Height decrease in the bundle	h_t	325 nm	*
Height of rootlet	h_r	975 nm	‡
Diameter of the stereocilia and kinocilium	d_{sc}	380 nm	†
Diameter at the cuticular plate of the stereocilia	d_r	125 nm	‡
Gap distance at the tip	g_t	10 nm	§
Gap distance at the bottom without taper	g_b	400 nm	§
Spacing of the stereocilia on the cuticular plate	s	780 nm	†
Angle between bundle shoulders	ϕ	96 °	†
Number of stereocilia along one edge	n_{edg}	5	†
Total number of stereocilia	n_{tot}	57	*

* Jacobs and Hudspeth (1990)
† This study
‡ Dr. A. J. Hudspeth (personal communication, November, 2009)
§ Estimated from micrographs

on this spacing, their height, and the angle of the bottom plane and the stereocilia axis at the tall and short edge. The diameter of the stereocilia needed to correctly reconstruct the geometry is $d_{sc} = 380$ nm. The kinocilium diameter measures about 320 nm including the outer membrane (Dr. A. J. Hudspeth, personal communication, November, 2009). Although the kinocilium diameter is less than the diameter of the stereocilia, for simplicity it is taken to be the same with $d_{sc} = 380$ nm.

The liquid coupling drag between stereocilia is inversely proportional to their wall-to-wall distance. For the model, a conservative estimate of the lower limit of this measurement was $s = 780$ nm for the center-to-center spacing of the stereocilia on the cuticular plate.* Thus the wall-to-wall distance at the bottom without tapering measures $g_b = 400$ nm with the given diameter of the stereocilia.†

The bundles seen from the top deviate from a perfect, equal-sided hexagonal arrangement (Figure 4.4). The angle between the bundle sides with the origin at the

---

*This also simplified the geometry construction of the full bundle.
†A closer distance at the bottom might be also due to a diameter increase of the stereocilia from the tip towards the upper end of the rootlet section (García et al. 1998).

kinocilium measures $\phi_{\text{dic}} = 102.6 \pm 5.9°$ ($n = 7$), which is less than 120° as for the equal-sided case. This implies an elongation of the hexagonal arrangement on the cuticular plate in the excitatory direction. In Figure 4.5 a scanning electron micrograph illustrates the hexagonal arrangement of the cuticular plate with removed stereocilia, which presents as well an elongation in the stimulus direction. For the model an angle of $\phi = 96°$ facilitated the construction.

In agreement with this observation are the stereocilia standing more parallel in the frontal view (Figure 4.3b) than in the side view (Figure 4.3c). In the side view the angle between the stereocilia axes from the tall and short edge measures about 30°, in the frontal view the angle between the axes of the edges is about 20°. For an equal-sided hexagonal arrangement, an angle in the side view of 30° would correspond to an angle of 26.1° in the frontal view.

The wall-to-wall distance at the tips is not known. Micrographs from transmission electron microscopy provide some rough estimations under the assumption that the top connector material survives the treatment for the electron microscopy. In Figure 4.1a the gap distance measures no more than $g_t = 10$ nm.

The geometrical dimensions used for the bundle model are given in Table 4.1 and illustrated in Figure 4.6. The kinociliary bulb was omitted from the model to simplify the liquid mesh around the kinocilium. It is assumed that potential increase in drag and inertia due to this spherical body are negligible compared to the overall bundle behavior. The values used for the model differ partly from the references, but allowed the construction of a parametric geometry and meshing with finite elements. Possible differences in the geometry do not critically affect the stiffness properties of the bundle as the Young's moduli of the material were adjusted by comparing the results provided by the finite element model to the experimentally measured stiffness.

4.2 Mechanical Properties

Two characteristic properties of the hair bundle from the bullfrog's sacculus are the stiffness and drag at the kinociliary bulb along the stimulus direction. A technique to measure the bundle stiffness is to attach a probe to the kinociliary bulb. Using this technique, Howard and Ashmore (1986) measured $k_{\text{hb}} = 256 \pm 28$ μN·m^{-1}. Two observations suggest that the bundle stiffness is related to bending of the stereocilia next to the cuticular plate. Firstly, the bundle stiffness is approximately proportional to the number of stereocilia. Secondly, the bundle stiffness is inversely proportional to the square of the height of application of the force measured from the cuticular plate. The second was observed as well for the hair bundles of turtles by Crawford and Fettiplace (1985). The concept of an elastically bending pivotal rootlet section and a rigid main part of the stereocilium was used in the model here as well. The rootlet section of the

4.2 Mechanical Properties

model consists of an elastic isotropic material whose Young's modulus is adjusted to reproduce measured bundle stiffness values.

Later studies by Howard and Hudspeth (1988) revealed an operation-point dependence of the hair-bundle stiffness. This dependence originates from the mechanoelectrical transduction channels which are connected to the tip links. The effective stiffness of the tip links was estimated as $k_{tl} = 500$ μN·m^{-1}. They computed this stiffness under the assumption that the system has a single degree of freedom.* As in the finite-element model a relative motion between stereocilia is in general possible, the tip link stiffness was readjusted here to match the experimental observations.

Cadherin 23, a part of the tip link, was analyzed in detail by Sotomayor et al. (2010). Their molecular dynamical simulations predict a stiffness of 40 mN·m^{-1} to 60 mN·m^{-1} for this part of the tip link, about hundred times stiffer than the experimental observations. Their *ab inito* calculations are missing the connections to the stereocilia bulk material as well as the other constituent of the tip links, the protocadherin 15, which might be more compliant.

Jaramillo and Hudspeth (1993) estimated the linear bundle stiffness with a displacement clamped probe as $k_{hb} = 650$ μN·m^{-1}. Later Martin et al. (2000) confirmed the values with $k_{hb} = 680 \pm 360$ μN·m^{-1}. Another technique to estimate the bundle properties is based on the observable Brownian motion of the hair bundle. Denk et al. (1989) retrieved by this method a stiffness of $k_{hb} = 350$ μN·m^{-1} and a bundle drag of $c_{hb} = 127$ nN·s·m^{-1}. Bashtanov et al. (2004) observed that removing all links from a chicken hair bundle except the top connectors reduces the bundle stiffness by 90 %. The stiffness of the hair bundle from the bullfrog's sacculus also reduces with the same treatment by a similar amount (Kozlov et al. 2011). The measured stiffness for the hair bundle without tip links is $k_{hb} = 19 \pm 12$ μN·m^{-1} ($n = 6$) and the drag is $c_{hb} = 30 \pm 13$ nN·s·m^{-1} ($n = 6$).

The kinocilium misses a distinct rootlet section and the inner structure is unchanged over the whole length. This suggests constant properties and a behavior as an Euler-Bernoulli beam characterized by its flexural rigidity. Okuno and Hiramoto (1979) measured for the cilia of echinoderm spermatozoa a flexural rigidity of 300 yN·m^2 to 700 yN·m^2. For the kinocilium of the hair bundle from the bullfrog's sacculus Kozlov et al. (2011) measured similar values with $EI_{kc} = 630 \pm 300$ yN·m^2 ($n = 13$). Besides the kinocilium individual stereocilia were analyzed. The drag of an individual stereocilium is with $c_{sc} = 16 \pm 5$ nN·s·m^{-1} ($n = 6$) about half of the drag of the whole bundle without tip links.

The properties of the top connectors and the kinociliary and basal links could not be determined experimentally so far. The properties have to be adjusted to observations where their function is involved and has a significant impact on the behavior. As the

---

*This also implies that the wall-to-wall distance is constant at the stereocilia tips or equivalently that the top connectors are infinitely stiff.

stereocilia are assumed to be rigid rods, the top connectors and basal links are mechanically coupled in parallel. The dense structures in the micrographs (Figure 4.1a) at the tips suggest a high stiffness for the top connectors. For the model, only the top connector stiffness is taken into account, the basal links being neglected. Also the lever arm is the longest for the top connectors with respect to the center of rotation. This further motivates the selection of the top connectors as the only elastic coupling element between stereocilia, besides the tip links.

4.3 Finite-Element Mesh

To analyze the mechanics of the hair bundle by means of the finite-element method, at first the geometry has to be provided. Based on the given geometry, the approximation of the domain by basis functions with local support can be accomplished by constructing the mesh (Section 3.1.1).

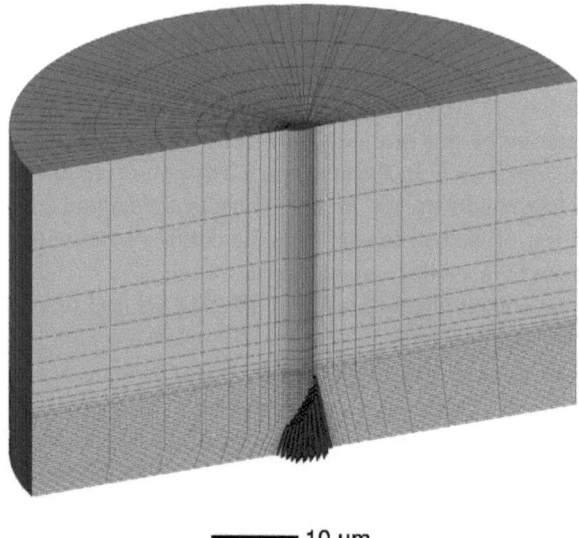

━━━ 10 µm

Figure 4.7: Finite-element mesh of the hair bundle model with a symmetry half of the surrounding liquid. For illustration purposes the symmetry half of the solid elements is mirrored here.

The geometry of the bundle is based on the dimensions given in Table 4.1. The bundle with a diameter of about 6 µm at the bottom and a maximum height of around 8.5 µm is placed into a cylinder of liquid with a diameter of 62 µm and a height of 37 µm on a flat, rigid surface (Figure 4.7). The excitations to the model are applied

4.3 Finite-Element Mesh

only in the symmetry plane along the physiological relevant direction. To reduce the model size, advantage is taken by regarding only one symmetrical half.

The tiny gap distance between stereocilia at their tips is less than a hundredth of the stereocilia height. The three dimensional configuration inhibit the correct arrangement of all cylindrical stereocilia with this gap distance to all next neighbors.[*] To circumvent this problem, the geometry was initialized with a slightly enlarged gap distance of 28 nm. Afterwards the local radius of the stereocilia was increased to ensure the correct wall-to-wall distance of 10 nm at the tips with a linear increase towards the bottom. The local radius of the stereocilia in each cross section between the points of closest approach was kept smooth as well. Different views of the closely apposed stereocilia are presented in Figure 4.8.

The mesh is multi-block structured and uses only hexahedral elements with quadratic shape functions (Chapter 3). The number of degrees of freedom is roughly 800,000 for the liquid and solid material together. The mesh is generated to resolve as accurately as possible the fluid mechanics between the stereocilia over lengths of 10 nm as well as the outer liquid drag with distances of more than 10 μm, while keeping the computation time as short as possible.[†] When possible, the element sizes vary smoothly with a factor of 1.4 from element to element.

A gap between two neighboring stereocilia is always filled by two elements (Figure 4.9). In a channel the fluid flow profiles are either linear or quadratic, if the thickness of the liquid layer is constant. The quadratic basis functions of the implemented finite element allow a correct solution for this configuration. Between two cylinders, the wall-to-wall distance variations are small in the vicinity of closest approach. For the drag force of two cylinders approaching each other, the mesh was fine enough to compute the reaction forces with an error that was less than a few percent (data not shown).

The top connectors and the tip links are incorporated as spring elements.[‡] To simplify the meshing, the element height is set to the geometry parameter h_t. The elastic spring elements can be placed at element nodes only. Thus the tip link's axis is at an angle of about 30° with respect to the stereociliary axis. The tip links are ending at the outer wall of the stereocilia. The top connector material spreads over some distance in vertical direction. To account for this, two springs are used, one next to the tips where the cylindrical part turns into a hemisphere and the other by the distance h_t lower. The top-connector stiffness is distributed equally between the upper and the lower spring.

---

[*]If the bundle is constructed by adding sequentially stereocilia to the model, new added stereocilia have to be placed at the correct gap distance to three next neighbors. This is an overdetermined problem. Increasing the gap distance sufficiently relieves this constrain for the slightly oblique stereocilia.

[†]The computation for single frequency on a single CPU took about 10 hours.

[‡]In Ansys (Version 11, Ansys Inc., Canonsburg, Pennsylvania) the Combin14 element is used. This element has only a longitudinal stiffness assigned. At the ends it couples just the translational degree of freedom. The rotations are uncoupled between this spring element and the solid element.

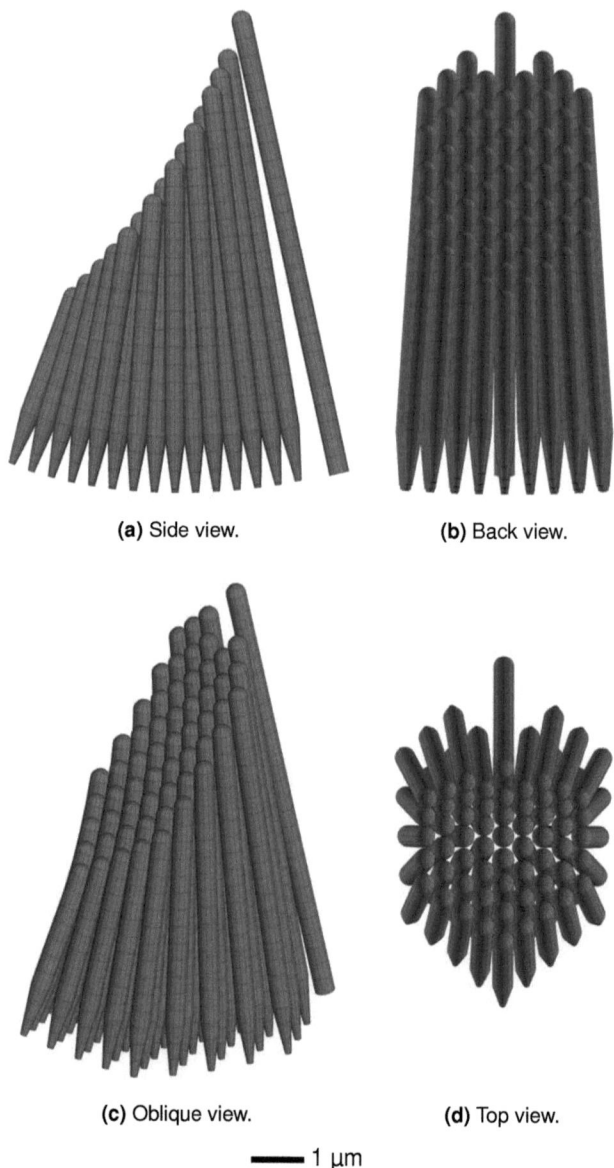

(a) Side view. (b) Back view.

(c) Oblique view. (d) Top view.

1 μm

Figure 4.8: Mesh of the finite-element hair-bundle model of the stereocilia and the kinocilium. For illustration purposes the solid elements were mirrored at the symmetry plane.

4.3 Finite-Element Mesh

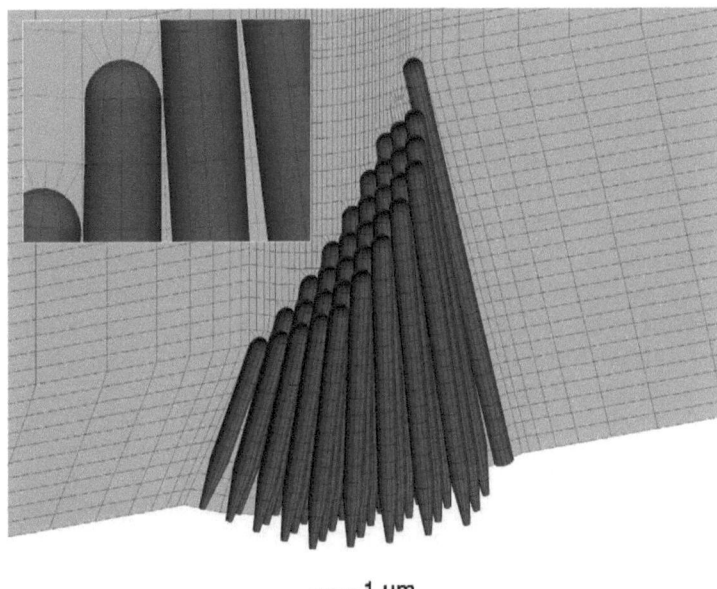

Figure 4.9: Mesh of the finite-element hair-bundle model with the mesh of the liquid shown in the symmetry plane. For illustration purposes the symmetrical half of the solid elements is mirrored here. The inset illustrates the gap near the tips.

The ends of these spring elements are inside the stereocilia on their axes. This was necessary to prevent unrealistic rotation of the stereocilia around their axis.

4.4 Material Properties

The material properties used in the model are given in Table 4.2. The properties for the inner ear liquids are discussed in detail in Section 2.2.2. The solid material properties are assumed to be isotropic (Section 2.2.1). The Poisson's ratio of all solid materials is set to $\nu = 0.3$, a typical value for elastic, isotropic material. The Young's modulus of the kinocilium is adjusted to reproduce the experimentally measured flexural rigidity of about 630 yN·m^2 (Kozlov et al. 2011). Based on the second moment of area for the cross section with radius $r = 0.19$ µm, the Young's modulus yields 615 kPa. In the finite-element model the equivalent flexural rigidity for a force at the kinociliary bulb differs due to the mesh size and three-dimensional effects. The Young's modulus is adjusted to compensate for this. The pivotal stiffness of the stereocilia is matched by the Young's modulus of the rootlet section to reproduce the experimentally measured bundle stiffness of 19 µN·m^{-1} (Kozlov et al. 2011). The pivotal stiffness of an individual stereocilium is about 10 aN·m·rad^{-1}. The Young's modulus of the rootlet section is also adjusted on the final mesh to compensate for possible mesh size dependencies. The Young's modulus assigned to the upper main part of the stereocilia is much higher than in the rootlet section to ensure a rigid motion of this upper part.

The stiffness of the kinociliary links was set to a high value to ensure a rigid connection. These links are just connecting the kinocilium to the stereocilia, but not the stereocilia to each other. The stiffness values for the tip links and top connectors were selected based on various computations in an evolutionary way.* The final results are in agreement with the various experimental observations discussed in Section 4.2. In the following Chapter 5 is the model used to analyze the importance of the elastic links between stereocilia in four different conditions: with only top connectors present, with only tip links present, with both types of links present simultaneously, and finally with no coupling links at all.

---

*With the various versions of the model 100,000 hours of computation time on single processors were used. An average computation with the model took about 10 hours yielding a total of about 10,000 computations.

4.4 Material Properties

Table 4.2: Material properties of the model of the hair bundle from the bullfrog's sacculus.

Name	Variable	Value	Ref.
Density of liquid and solid	ϱ	1 Mg·m^{-3}	*
Viscosity of liquid	η	1 mPa·s	*
Bulk modulus of liquid	K	1 GPa	*
Young's modulus of rootlet	E_r	111 kPa	†
Young's modulus of stereocilium	E_{sc}	1 GPa	‡
Young's modulus of kinocilium	E_{kc}	585 kPa	†
Poisson's ratio of solid material	ν	0.3	§
Stiffness of tip link	k_{tl}	1 mN·m^{-1}	†
Stiffness of top connector	k_{tc}	20 mN·m^{-1}	†
Stiffness of kinociliary link	k_{kl}	1 N·m^{-1}	‡

* Similar properties to water (Section 2.2.2)
† This study
‡ Estimated as rigid
§ Elastic isotropic material

Chapter 5

Mechanics of the Hair Bundle from the Bullfrog's Sacculus

The finite-element model presented in Chapter 4 allows a detailed analysis of the mechanics of the hair bundle, by studying the bundle response to different load conditions. Three different modes of excitation are investigated here:

Kinociliary Bulb Displaced The response to a displacement at the kinociliary bulb mimics the excitation of the bullfrog hair bundle under *in vivo* conditions.

Stochastic thermal excitation To understand the mechanical coupling between the stereocilia the coherency is evaluated between individual stereocilia in the bundle for forcing by a thermal bath.

Tip-Link Force A force applied along the tip links is used to investigate response of the bundle due to non-linear forces generated by channel gating.

The observation and excitatory direction always align with the stimulus direction of the bundle. In all cases the bottom plate is fixed and rigid. The arbitrary outer boundary of the liquid domain moves freely and no forces are applied there. The sufficient size of the domain is validated as the dimensions of the fluid compartment are larger than a millimeter. The liquid around the bundle is modeled without taking into account nearby bundles or the otolithic membrane. The range of frequencies from 1 mHz to 100 kHz encompasses more than the complete relevant range of physiological conditions and experimentally available data. For the computed frequencies the mechanics are dominated by stiffness, drag, and inertial forces.

The properties and consequences of the elastic linkages between individual stereocilia are studied in detail. Four different configurations are computed for the elastic links in the bundle between stereocilia:

All links As in an intact hair bundle there are both types of elastic links in the bundle: the top connectors and the tip links. Most data are available for this configuration.

Tip links Only elastic tip links connect the stereocilia. So far no treatment exists which removes the coupling top connectors, but not the tip links.

Top connectors Only top connectors link the stereocilia. By treatment with BAPTA the tip links can be removed from the bundle.

No links There are no elastic links between stereocilia. So far no treatment exists which removes the coupling top connectors and the tip links.

In all cases the viscous liquid between and around the stereocilia is present and the kinociliary links tightly couple the kinocilium to the next neighboring stereocilia.

In this chapter no parameter studies are presented. The properties of the model were carefully selected and adjusted in numerous computations. In the Appendix C parameters are presented with variations of the wall-to-wall distance at the tips, of the top-connector stiffness, of the kinocilium stiffness, of the stereocilia rootlet stiffness, of the density, and the viscosity between the stereocilia. These parameter studies show how the individual observed quantities are affected by the selected properties.

5.1 Kinociliary Bulb Displaced

The kinociliary bulb is the connecting structure between the otolithic membrane and the hair bundle. The response of the bundle to this stimulus is investigated by the frequency dependent displacement patterns. Time traces of individual stereocilia tips are given and the force at the kinociliary bulb in the stimulus direction, which is the physiological relevant direction, are analyzed.

In Figure 5.1 the motion patterns in response to the displacement of the kinocilium are presented for a cycle of the sinusoidal excitation signal. The amplitudes are scaled by the individual stereocilia height to plot with the same amplitude stereocilia moving with the same pivotal rotation angle. The phase relation of the individual markers is given in Figure 5.1e.

A bundle with just viscous liquid filling the gap has two regimes (Figure 5.1a). Below 0.1 Hz the drag is dominated by the squeezing between the displaced stereocilia and the next neighbors. Just the elastically coupled next neighbors move with the kinocilium. The viscous forces at these low frequencies are not large enough to work against the individual, pivotal stiffness of the stereocilia. With increasing frequency the viscous forces increase relatively to the elastic forces of the stereocilia and more and more stereocilia move together with the kinocilium. Around 0.1 kHz the bundle starts to move coherently with no noticeable relative motion between the stereocilia. At very high frequencies such as 100 kHz inertial forces come into play and reduce the amplitude of those stereocilia which are further away from the point of excitation, the kinocilium. Adding the stiff top-connector material blocks relative motions in the bundle (Figure 5.1b). For all frequencies the bundle moves as one unit, except the high frequencies where again the inertial forces intervene. Adding tip links to the model introduces some elastic coupling between the stereocilia (Figure 5.1c). Only stereocilia

5.1 Kinociliary Bulb Displaced

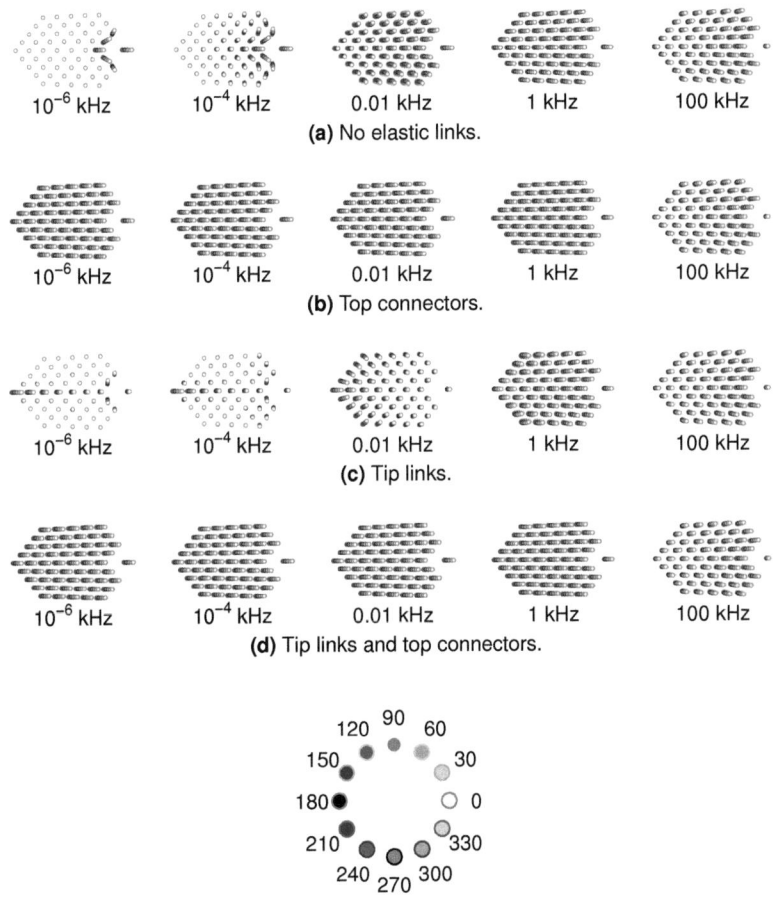

Figure 5.1: Time traces of a cycle of the stereocilia tips for a displacement prescribed at the kinocilium and phase relation of the time traces with time advancing during the cycle clockwise. The displacements are scaled arbitrarily and divided by the individual stereocilia height for illustration purposes. The time traces are centered around the location of the rootlets.

aligned along the stimulus direction are coupled by the tip links. The viscous coupling is too weak to suppress the relative motion at low frequencies. For a deflection in the excitatory direction the pivoting stereocilia are pulled closer to each other by the oblique tip links, which amplifies the motion at the short edge. If tip links and top connectors are present the top connectors dominate the motion pattern (Figure 5.1d) and the bundle moves without noticeable relative motion for all frequencies. Also for the last two cases the motion pattern at 100 kHz is not affected by the additional elastic links, indicating that the problem is governed by inertial and viscous forces.

The mechanical behavior is further characterized by the ratio of the reaction force at the kinociliary bulb to kinematic variables. There are three possible kinematic variables, namely displacement, velocity, and acceleration. The force with respect to them depends upon the frequency. At low frequencies the motion is quasi-static and all relaxations have enough time to equilibrate, thus damping forces are small. Inertial forces scale with the frequency to the power of two and decay faster with decreasing frequency than damping forces and therefore play a minor role at low frequencies. The hair bundle is overdamped and thus viscous forces gain importance smoothly with increasing frequency. For higher frequencies the inertial forces dominate and delay the motion by a half cycle, if no other modes of oscillations play a role. This is the case here.

Assuming a mass-spring-damper system with a single degree of freedom allows us to characterize the dynamics of the hair bundle. Denk et al. (1989) measured for an intact bundle a stiffness of $k_{hb} = 350$ μN·m^{-1} and a drag coefficient of $c_{hb} = 127$ nN·s·m^{-1}. The effective mass for a moving bundle is not only the mass of the stereocilia. The liquid in between the stereocilia and the entrained external liquid have to be taken into account as well. The bundle's overall mass is estimated as roughly three times the mass of the elastic structures $m_{hb} = 0.1$ ng. A more rigorous estimation of this effective mass will be given further below in Section 5.1.3. The undamped resonance frequency of the oscillator is

$$f_0 = \frac{1}{2\pi} \sqrt{\frac{k}{m}}. \tag{5.1}$$

With the value for an intact hair bundle this yields $f_0 = 9.4$ kHz. The damping ratio

$$\zeta = \frac{c}{2\sqrt{km}} \tag{5.2}$$

is larger than 1 if the system is overdamped. With the above given values $\zeta = 11$ the bundle is clearly overdamped. The experimental data for a hair bundle without tip links provide a stiffness of $k_{hb} = 19$ μN·m^{-1} and a drag of $c_{hb} = 30$ nN·s·m^{-1}. Assuming as above an effective mass of $m_{hb} = 0.1$ ng decreases the undamped resonance frequency

5.1 Kinociliary Bulb Displaced

to 2.2 kHz, while the resulting damping ratio of $\zeta = 10.9$ indicates overdamping also in this situation.*

A way to analyze the ratio of force and a kinematic variable over frequency in more detail is by the impedance, which is the complex ratio of force divided by velocity.[†] The overdamped hair bundle exhibits a smooth variation of the forces and motions in the range of interest. Under these specific circumstances the real part of the ratio force to displacement, to velocity, and to acceleration is more descriptive. Therefore, the computational results are presented as drag, stiffness, and inertial coefficients, which can be directly compared to experimental observations.

5.1.1 Drag

The real part of the ratio of force divided by the velocity provides the drag coefficient

$$c_{hb}(f) = \text{Re}\left(\frac{F_{kb}(f)}{v_{kb}(f)}\right), \tag{5.3}$$

which is a function of the frequency f based on the force F_{kb} and velocity v_{kb}. Both are evaluated along the excitatory direction at the kinociliary bulb. $\text{Re}(\cdot)$ is the real value operator. There are no forces counteracting the drag in the model.

Figure 5.2 presents the drag coefficient of the bundle. The purely liquid-coupled bundle has a drag coefficient of up to 4,400 nN·s·m^{-1} at a frequency of 1 mHz. For these high drag values the presence of the external liquid does not alter the results. The drag coefficient decays as the coupling forces of the viscous liquid between the stereocilia overcome the pivotal stiffness of the stereocilia. For frequencies of 0.1 kHz and higher the drag is around 50 nN·s·m^{-1} with the surrounding liquid and around 13 nN·s·m^{-1} without. This indicates that the coherent bundle drag is mainly dominated by the external liquid. Above 10 kHz the drag increases again due to the decrease of the thickness of the oscillatory boundary layer, as for an oscillating sphere (Section 3.2.1).

The top connectors fully block the relative modes and the associated drag. The value is constant around 55 nN·s·m^{-1} and again slightly increases from around 10 kHz. Again this is mainly dominated by the surrounding liquid, as the results without yield

---

*The overdamping implies that no distinct peaks exist in the amplitude spectra. Therefore no resonance frequency can be identified from the bundle response. The cross-over frequency where the dominant forces change from stiffness to inertia still exists and has a characteristic frequency assigned. In this work this frequency is labeled as *undamped resonance frequency*.

[†]The impedance is widely used in acoustics, electrical circuits, optics and other fields where several resonance frequencies play a role.

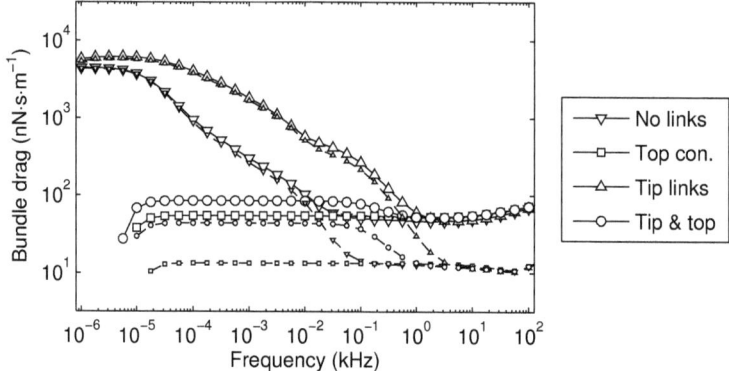

Figure 5.2: Drag of the hair bundle computed at the kinociliary bulb from the real part of the applied force divided by the velocity in the stimulus direction. The solid lines with large markers are for the bundle with surrounding liquid. The dashed lines with small markers are for the bundle without surrounding liquid. Stereocilia are always coupled by the viscous liquid and elastically coupled only by horizontal top connectors (Top con.), tip links (Tip links), both types of elastic links (Tip & top), and no elastic links (No links).

a value of around 13 nN·s·m^{-1}. Below 10 mHz the drag evaluation fails probably due to numerical inaccuracies.[‡]

A bundle with only tip links as elastic coupling elements between stereocilia exhibits highest damping compared to the other three cases. Below the frequency of 20 mHz the drag is around 6,000 nN·s·m^{-1}. With increasing frequency the drag is always higher than the drag of the purely liquid-coupled bundle. The oblique tip links transfer about the same displacement amplitude from the tall stereocilium to the next shorter one, but the lever arms of the center of the pivotal motion with respect to tip-link connection point differ. The shorter stereocilium displaces at a lower height and therefore rotates further causing additional relative motion with associated drag. From around 3 kHz to higher frequencies the viscous coupling of the liquid overcomes the tip-link stiffness and the drag aligns with the purely liquid coupled solution. Starting from around 3 kHz the drag is, as in the previous cases, about 50 nN·s·m^{-1} with the external liquid and around 13 nN·s·m^{-1} without.

The drag of a bundle with both types of elastic links between stereocilia follows closely the drag of a bundle with just top connectors for frequencies above 1 kHz. The same issues with the drag evaluation are present for very low frequencies. For frequencies from 0.1 Hz to 0.1 kHz the relative motion induced by the tip-links increases the

---

[‡]The computations are performed with complex numbers. For the low drag cases the ratio of the real to the imaginary part is about 10^6. The very small imaginary part is needed for the drag evaluation. No further studies investigated this issue in this work. The frequencies where the drag evaluation fails are below the frequency where the quasi-static response is reached.

drag to 85 nN·s·m^{-1} with external liquid and to 43 nN·s·m^{-1} without. This indicates that the displacement patterns of the relative modes propagate in the external liquid. This higher drag at low frequencies is in agreement with the experimental observation by Denk et al. (1989) that a healthy bundle has a higher drag than a solid object of the same dimensions.

Drag by the External Liquid

The drag induced by the external liquid is further analyzed on simplified geometries. Denk et al. (1989) used as reference a sphere in an unbounded liquid in steady motion. Based on Stokes's analytical solution the associated drag reads

$$c_{\text{sph}} = 6\pi\eta r \tag{5.4}$$

with the sphere radius r and the dynamic viscosity η. For a viscosity of $\eta = 1$ mPa·s and an average radius of $r = 2.5$ μm the drag coefficient yields $c_{\text{sph}} = 47.1$ nN·s·m^{-1}, a value similar to the bundle computations with only top connectors present (Figure 5.2).

The hair bundle is not translating, but pivots around the base. To estimate the drag of a pivotal motion the drag of a rotating semi-ellipsoid is analyzed. The ellipsoid elongates along the major axis and has equal minor axes. The plane of the minor axes divides the ellipsoid and mimics the bottom plane of the bundle. The tip of this ellipsoid corresponds to the tip of the kinocilium. The drag coefficient is computed based on the force required at the tip to rotate the semi-ellipsoid around one of the minor axes. The velocity is evaluated at the same location. Based on the analytical solution of Perrin (1934) the drag coefficient reads

$$c_{\text{ell}} = \frac{8}{3}\pi\eta \frac{\sqrt{h^4 - h^2 r^2}\left(h^4 - r^4\right)}{\left((2h^2 - r^2)\operatorname{arccosh}\left(\frac{h}{r}\right) - \sqrt{h^4 - h^2 r^2}\right)h^3} \tag{5.5}$$

with the semi-ellipsoid height h and minor axis radius r. The same radius and viscosity are used for an estimation of the drag coefficient as for the sphere. The height is assumed to be similar to the kinocilium height of the model with $h = 8$ μm. The drag of this pivoting semi-ellipsoid yields $c_{\text{ell}} = 24.9$ nN·s·m^{-1} and is about half of the drag estimated for the sphere. An ellipsoid is very smooth and without sharp corners, whereas the hair bundle instead ends at the tips of the stereocilia, which are rounded with their radius. The drag of the ellipsoid is therefore a lower estimate. In agreement with this estimation is the experimentally measured drag of 30 ± 13 nN·s·m^{-1} for a bundle without tip links (Kozlov et al. 2011). This confirms that the external liquid drag dominates the drag for a bundle without tip links. The remaining drag due to internal relative sliding between stereocilia is smaller.

For an individual stereocilium the drag is about half that of the full bundle with 60 stereocilia. The experimental data provide a value of 16 ± 5 nN·s·m^{-1} (Kozlov et al. 2011). This value can be compared with the drag of a pivoting cylinder. As for the equivalent ellipsoid the geometry is simplified to a cylinder with the height h and radius of a stereocilium r. The axis of rotation lies in the bottom plane of the cylinder and the drag is again evaluated at the tip surface center. Based on the approximative solution of Broersma (1960) with the corrected coefficients of Broersma (1981) the drag coefficient reads

$$c_{\text{cyl}} = \frac{4}{3} \pi \eta \frac{h}{\sigma - 1.14 - 0.2\,\sigma^{-1} - 16\,\sigma^{-2} + 63\,\sigma^{-3} - 62\,\sigma^{-4}} \quad (5.6)$$

with

$$\sigma = \ln\left(\frac{2h}{r}\right) \quad \text{and} \quad 0 \leq \frac{1}{\sigma} < 0.45 \,. \quad (5.7)$$

For a stereocilium with a height $h = 8$ μm, a radius $r = 0.19$ μm and the dynamic viscosity $\eta = 1$ mPa·s the equivalent drag of a pivoting cylinder displaced at the tip yields $c_{\text{cyl}} = 11.2$ nN·s·m^{-1}. The estimation based on the semi-ellipsoid fitted into the cylinder results in 8.5 nN·s·m^{-1}. This is a smaller value, as the drag of the cylinder combines the drag of an ellipsoid and the drag by the end surfaces (Broersma 1960).

A finite-element model of a pivoting cylinder is used for two purposes: at first for the validation of the size of the external liquid domain to ensure that the arbitrary outer boundary does not alter the results and secondly to compare the finite-element results with the above given analytical estimations. All relevant results are collocated in Figure 5.3.

The finite-element mesh corresponds to the mesh of the bundle model for the external liquid. The bundle is replaced by a cylindrical volume with a height of 8 μm. The variable radius encompasses the geometry of a single stereocilium and the whole hair bundle. The cylinder is displaced with a linear velocity profile over the height in a single direction parallel to the bottom plane. The motion in axial direction on the cylindrical wall is set to zero. The reaction moment of torque is computed with respect to the bottom center of the cylinder. The division by the cylinder height and the velocity at the cylinder tip provides an equivalent drag coefficient from the finite-element computation.

The finite-element solutions are given for two situations. The outer liquid boundary is either held or free to move. The later case corresponds to the case used for all hair bundle computations in this work. The liquid properties are the same as those given in Table 4.2. The frequency in the computations is 1 Hz. The characteristic dissipation length for the liquid at this frequency is about 400 μm, which is much larger than the computational domain. Therefore the problem is quasi-static. Both finite-element

5.1 Kinociliary Bulb Displaced

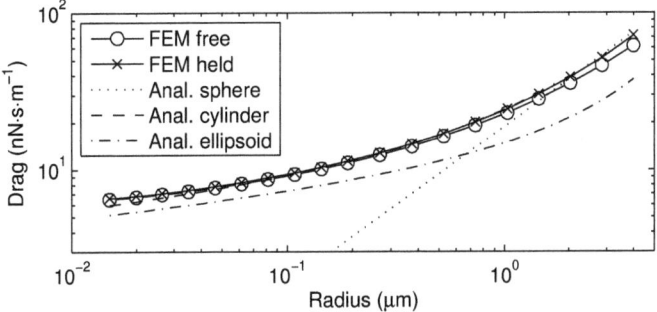

Figure 5.3: Drag of a cylinder displaced by a linear profile over the height on a fixed plate. The height is $h = 8$ µm. The drag is given for an equivalent force at the cylinder tip. For the finite-element results the liquid domain has an outer diameter of 62 µm and a height of 37 µm. The outer boundary is either free to move (FEM free) or held in position (FEM held). As comparison the analytical solution of a translating sphere (anal. sphere), of a rotating half ellipsoid (Anal. ellipsoid) based on the work from Perrin (1934), of a translating sphere (Anal. sphere), and the fitted function for a cylinder (Anal. cylinder) from Broersma (1981) are given. The later is only given up to the valid range of $r < 0.22\,h$.

solutions align well with each other. For a too thin outer boundary layer they would differ significantly as presented in the example of an oscillating sphere in Section 3.2.1. This implies that the external liquid domain is sufficiently large for the hair bundle model for frequencies down to 1 Hz.

The finite-element solution aligns as well with the analytical approximation of the cylinder drag. This suggests a minor importance of the rigid bottom plane where the bundle inserts. The ellipsoid fitting into the cylinder has a lower drag, as the volume is smaller and the end surfaces are missing. The variation of the drag versus the radius is similar, whereas the translating sphere has a different variation. For radii similar to the height of the cylinder the values are comparable, but for small radii the drag of the sphere is much less, as the drag is determined by the largest geometrical dimension of the object (Howard 2001).

Drag in the Bundle

To understand the liquid coupling between the stereocilia and the losses in the bundle in more detail, the interaction of two pivoting cylinders is investigated. There are four independent modes of motion:

Squeezing ($\rightarrow\leftarrow$) Both cylinders pivot towards their common center. This is the relative mode in a bundle induced by the oblique tip links for example.

Shearing ($\uparrow\downarrow$) The cylinders pivot in opposite directions and normal to the plane containing their axes. This mode represents shearing in the bundle.

Transversal (↑↑) The cylinders pivot in the same direction normal to the plane containing their axes. This mode represents for example the motion of neighboring stereocilia of the symmetry plane moving normal to this plane and so also normal to the stimulus direction.

Sliding (→→) The cylinders pivot in the same direction and in the plane containing their axes. This mode represent for example the motion of neighboring stereocilia of the symmetry plane moving in the stimulus direction. This is the physiologically relevant mode in a bundle moving coherently for stereocilia standing behind each other.

In all situations it is assumed that the amplitudes of motion are the same for the two cylinders.

The assigned drag to all four modes of motion is computed with a finite-element model of two cylindrical bodies with dimensions corresponding to the ones of the stereocilia in the bundle. The height measures 8 μm and the diameter 380 nm. The wall-to-wall distance at the bottom is 400 nm and at the tips values from 5 nm to 150 nm are assigned. There is no liquid above the cylinders and so their caps do not induce any drag. The external liquid layer around the cylinders has a thickness of about seven times the diameter, which is sufficiently large to represent an infinite external liquid radius. The displacement profiles applied are similar to the single cylinder computations from the previous section with the exception that displacements along the axis are applied here as well. They are set in such a way that the cylinders rotate like solid objects.

The liquid properties are given in Table 4.2. The frequency of the computations is 1 Hz. The velocities are set along the cylinder wall without an axial component as for the single cylinder described above. The equivalent reaction forces are evaluated based on a momentum balanced with respect to the center of rotation.

The squeezing mode of motion induces the highest drag. In the gap between the cylinders the velocity profiles over the gap height are quadratic with a large velocity gradient next to the wall. Together with Dr. Thomas Risler (Institut Curie, Physical Chemistry, Paris) an analytical approximation was derived for small gaps with the minimum gap distance varying over the cylinders length smoothly. For a relative motion between the two cylinders this reads

$$c_{\text{squeeze}} = \pi \eta h \frac{\xi^2 (3 + \xi) \chi^3}{(1 + \xi)^3} \tag{5.8}$$

where

$$\xi = \sqrt{\frac{g_t}{g_b}} \quad \text{and} \quad \chi = \sqrt{\frac{r}{g_t}}. \tag{5.9}$$

5.1 Kinociliary Bulb Displaced

Here h is the cylinders height, r the cylinders radius, g_t and g_b the wall-to-wall distance at the tip and bottom, and η the dynamic viscosity. A similar drag coefficient estimation was conducted by Zetes (1995) for parallel cylinders. Further details are given in the Appendix B.1.

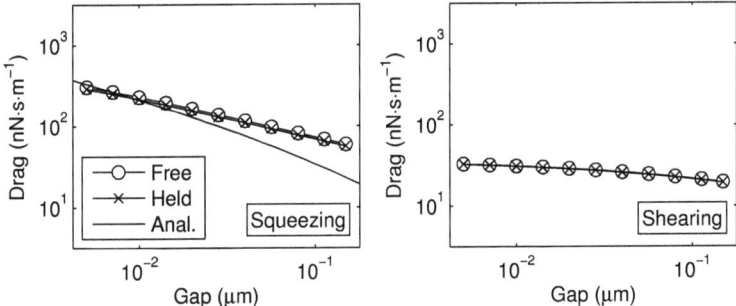

Figure 5.4: Drag coefficient of pivoting cylinders moving in the opposite directions in squeezing and shearing modes. The finite-element results are for a fluid free to move in all directions (Free) and for a two-dimensional constraint where the flow in the cylinders mean axial direction is held (Held). For comparison the analytical approximation (5.8) is given (Anal.) for the squeezing mode.

In Figure 5.4 the analytical approximation is compared with the finite-element solution. They converge to the same solution for small gaps. For comparison purpose liquid motions are also blocked in the axial direction to enforce a two-dimensional flow pattern. This keeps the resulting drag coefficient marginally affected. If the gap at the tips decreases to zero the drag coefficient diverges. Even for almost parallel cylinders the drag is significantly higher than for a single cylinder with a pivotal motion, which was estimated and computed to be in the order of 10 nN·s·m^{-1}. For the shearing mode of motion no analytical estimation could be derived and found in the literature.[*] The finite-element results provide lower drag values than for the squeezing mode and lower divergence for the decreasing gap size.

For the modes of motion with the same orientation, the transversal and the sliding mode, the drag values are significantly lower than for the relative modes (Figure 5.4) and even below the drag of a single cylinder in the liquid. The reason here is the missing cylinder ends. To check the influence by the external liquid, the layer thickness of the liquid around the cylinders was decreased from seven times the diameter to just the diameter. The drag decreases for both modes from 1.4 nN·s·m^{-1} to 0.12 nN·s·m^{-1} indicating that the shearing of the liquid between the cylinders is negligible compared to the single cylinder drag. This agrees also with the observation that the overall bun-

---

[*]The work of Jeffrey and Onishi (1981) uses bipolar coordinates for viscous flow of a cylinder next to a wall. With some modifications this might provide a general solution.

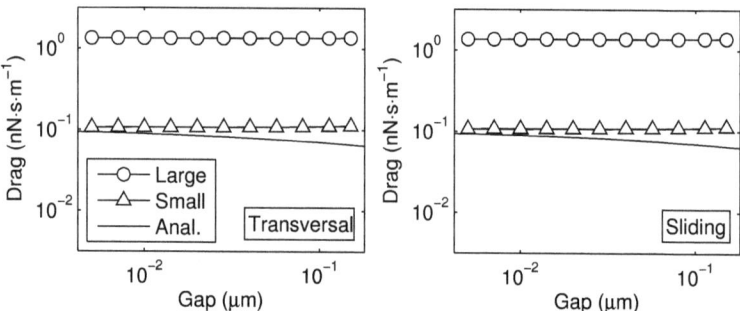

Figure 5.5: Drag coefficient of pivoting cylinders moving in the same direction in transversal and sliding modes. The finite-element results are for a large radius of the volume of the external liquid (Large) and a small radius (Small). For comparison an analytical approximation is given (Anal.).

dle drag of a coherent bundle without external liquid is 10 nN·s·m^{-1} (Figure 5.2) as there are roughly 50 pairs of sliding stereocilia, which yields a sliding drag per pair of 0.2 nN·s·m^{-1}. The comparison with an analytically based estimation supports this low sliding drag. The algorithm to compute this estimation is described in detail in Appendix B.2.

5.1.2 Stiffness

Like the drag coefficient of the bundle (5.3), the ratio of the force divided by the displacement characterizes the stiffness provided that inertial forces are not dominant. The stiffness coefficient

$$k_{\text{hb}}(f) = \text{Re}\left(\frac{F_{\text{kb}}(f)}{u_{\text{kb}}(f)}\right) \qquad (5.10)$$

is a function of frequency f based on the force F_{kb} and displacement u_{kb}. Both are evaluated along the excitatory direction at the kinociliary bulb. This ratio is in effect the subtraction of the inertial force from the stiffness force. This ratio yields zero at the undamped resonance frequency as the stiffness and inertial forces are equal and cancel each other. For higher frequencies this ratio becomes negative, as long as there are no other oscillatory modes gaining importance.

Figure 5.6 presents the bundle stiffness as a function of frequency. For the bundle without any elastic coupling between stereocilia the low-frequency response is set by the stiffness of the kinocilium and the tightly connected next neighbors to a value of 7.4 μN·m^{-1}. For the frequencies from 1 Hz to 10 Hz the liquid coupling intervenes and the stiffness increases to 20.7 μN·m^{-1}, corresponding to the sum of the stiffness of all stereocilia and the kinocilium.

5.1 Kinociliary Bulb Displaced

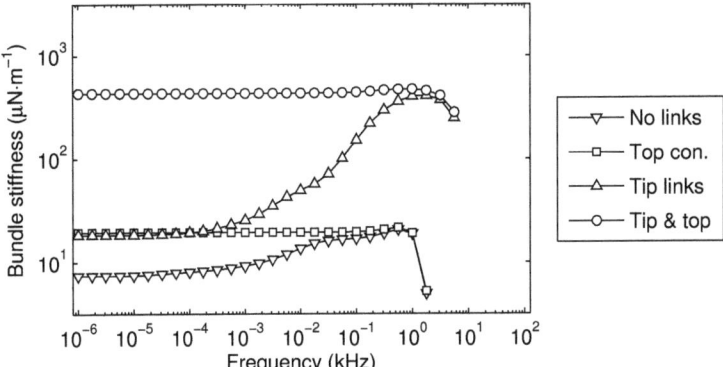

Figure 5.6: Stiffness of the hair bundle computed at the kinociliary bulb from the real part of the ratio force divided by displacement in the stimulus direction. Stereocilia are always coupled by the viscous liquid and elastically coupled only by horizontal top connectors (Top con.), tip links (Tip links), both types of elastic links (Tip & top), and no elastic links (No links).

The stiffness of a single stereocilium k_{sc} at the height h can be estimated by the pivotal stiffness k_{rot} based on the conservation of momentum with respect to the center of rotation as

$$k_{sc} = \frac{k_{rot}}{h^2}. \tag{5.11}$$

The experimentally determined pivotal stiffness is 10 aN·m·rad^{-1}. The sum over all stereocilia in the bundle yields 13.5 µN·m^{-1}. This is the contribution of the individual stereocilia to the overall bundle stiffness. The bending stiffness of the kinocilium converts into a stiffness by

$$k_{kc} = \frac{3\,EI}{h^3} \tag{5.12}$$

assuming a single force at h acting on a clamped Euler-Bernoulli beam with flexural rigidity EI. With the experimentally determined flexural rigidity of 630 yN·m^2 this yields a stiffness contribution by the kinocilium of $k_{kc} = 6.9$ µN·m^{-1}. The sum of all stereocilia and the kinocilium stiffness is 20.4 µN·m^{-1}, which is close to the value given above of 20.7 µN·m^{-1}. Thus, the individual stiffness properties are correctly assigned to the model. The undamped resonance frequency of the bundle without elastic links between stereocilia occurs around 2 kHz.* Adding the top connectors to the bundle prevents a decoupling of the stereocilia. The stiffness remains constant with about

---

*At the undamped resonance frequency the coefficient $k_{hb} = 0$. With the logarithmic scaling of the ordinate k_{hb} can be displayed only for frequencies below resonance.

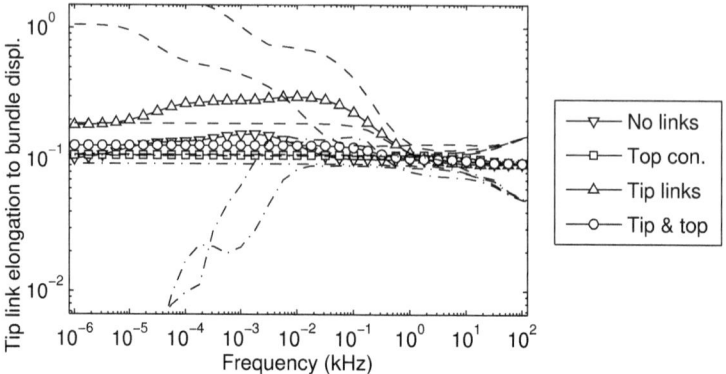

Figure 5.7: Length change of tip links divided by the bundle displacement at the kinociliary bulb. The solid lines with the markers are the mean values, the dashed lines the maximum values, and the dashed-dotted lines the minimum values of all tip links. Stereocilia are always coupled by the viscous liquid and elastically coupled only by horizontal top connectors (Top con.), tip links (Tip links), both types of elastic links (Tip & top), and no elastic links (No links).

20 μN·m^{-1} for all frequencies below 0.1 kHz as all stereocilia are equally deflected. The addition of tip links increases the bundle stiffness to 440 μN·m^{-1}. With this increased stiffness the undamped resonance shifts up to 6 kHz. The bundle equipped with only tip links between stereocilia has a stiffness of 18.5 μN·m^{-1} for low frequencies. Around medium frequencies the viscous coupling is strong enough to couple some stereocilia and the stiffness increases with frequency. At 1 kHz the viscous coupling between the stereocilia overcomes the relative motion induced by the tip links. Henceforth the tip links are almost fully coupled to the bundle motion and the bundle stiffness reaches a value of 410 μN·m^{-1}. For frequencies higher than the undamped resonance frequency the ratio k_{hb} is negative and decreases monotonically up to 100 kHz proportional to the square of frequency for all configurations (data not shown). This indicates that no further resonance frequencies play a role for an excitation on the kinociliary bulb in the direction of sensitivity for the investigated range of frequencies.

The length change of the tip links with respect to the bundle displacement accounts for the tip-link stiffness contribution to the overall bundle stiffness (Jacobs and Hudspeth 1990). The length change of the individual tip-link distance with respect to the bundle displacement is shown in Figure 5.7. In the model each tip-link strains individually and frequency-dependent. To reduce the amount of data shown in the plot only the minimum, mean, and maximum values are presented. If tip links are not present, as for the cases with just top connectors and no elastic links, the values correspond to the length change of a virtual link with zero stiffness at the same position. The top connectors hold the bundle tightly together up to a frequency of 1 kHz. All tip links

5.1 Kinociliary Bulb Displaced

elongate by a similar value, ranging from 0.106 to 0.109. For higher frequencies the bundle moves with increasing relative motion due to the inertial and viscous forces, independent of the elastic linkage. If the tip links are added to the model, the frequency response remains flat up to 0.1 kHz. The mean value is increased to 0.13 and the values have a broader range from 0.09 to 0.19. At higher frequencies the viscous coupling holds the stereocilia more together and the values asymptotically align with the results from the bundle with just top connectors. Removing all elastic links causes different individual relative motion in the bundle with a mean value still around 0.1. Just for 1 kHz and higher frequencies the values converge to a common narrow range as the liquid coupling starts to dominate. The bundle with only tip links has a qualitatively comparable behavior. For frequencies below 1 kHz the elongation is larger due to the oblique orientation of the tip links and the splay they induce.

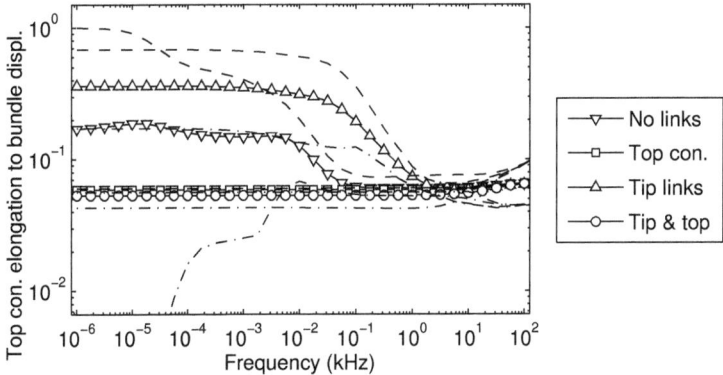

Figure 5.8: Length change of top connectors divided by the bundle displacement at the kinociliary bulb. The solid lines with the markers are the mean values, the dashed lines the maximum values, and the dashed-dotted lines the minimum values of all tip links. Stereocilia are always coupled by the viscous liquid and elastically coupled only by horizontal top connectors (Top con.), tip links (Tip links), both types of elastic links (Tip & top), and no elastic links (No links).

Figure 5.8 presents the length change of the top connectors divided by the displacement at the kinociliary bulb. In the model the top connectors link the individual stereocilia axes directly and so this is a measure of the relative motion of the stereocilia. For the bundle with all links the value is around 0.05 for frequencies below 3 kHz. If the tip links are removed the scatter decreases and the mean value becomes 0.06 for the same frequency range. In both situations the mean value increases slightly up to 0.065 at 100 kHz and the scatter increases again with values from 0.04 to 0.09. Without any elastic coupling the relative motion between the stereocilia is roughly 0.15 up to 10 Hz where viscous coupling starts to play a role. For the bundle with just tip links the mean relative motion is larger at low frequencies with a mean value of 0.35. Around 150 Hz the relative motion decreases and aligns with the other link configurations for higher frequencies.

This analysis illustrates the importance of elastic coupling for frequencies below 100 Hz. For higher frequencies the viscous forces start to dominate until the inertial forces prevail at even higher frequencies.

5.1.3 Inertia

Although of minor importance for physiologically relevant frequencies, the kinematic variable acceleration can serve as a reference for the force experienced by the hair bundle at the kinociliary bulb. Similar to the drag (5.3) and the stiffness (5.10) coefficients of the bundle, the coefficient

$$m_{hb}(f) = \text{Re}\left(\frac{F_{kb}(f)}{a_{kb}(f)}\right) \tag{5.13}$$

is defined with units of mass as a function of frequency f. The force F_{kb} and acceleration a_{kb} are both evaluated along the excitatory direction at the kinociliary bulb. This coefficient has the opposite sign of the stiffness coefficient (5.10) and is positive above the undamped resonance frequency.

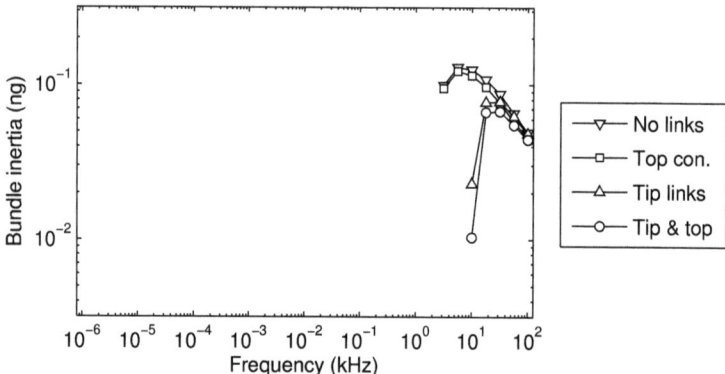

Figure 5.9: Inertia of the hair bundle computed at the kinociliary bulb from the real part of the ratio force divided by acceleration in the stimulus direction. Stereocilia are always coupled by the viscous liquid and elastically coupled only by horizontal top connectors (Top con.), tip links (Tip links), both types of elastic links (Tip & top), and no elastic links (No links).

Figure 5.9 presents the inertia of the hair bundle. For this a reference value can be defined as the one for the elastic volumetric materials. From the finite-element model this yields 0.037 ng. If the fluid in between the stereocilia is added this mass becomes

0.050 ng. The bundle without any elastic coupling has a maximum inertia mass of 0.13 ng around 6 kHz and decays from there. The corresponding value of the bundle with only top connectors is a little bit less with 0.12 ng. Both values are more than twice the bundle's mass. The additional mass originates from the liquid moving together with the bundle. With increasing frequency the viscous boundary layer thickness decreases and so does the liquid mass moving with the bundle. If tip links are present in the bundle the undamped resonance frequency is higher, as mentioned above. The inertial coefficient is positive starting from 10 kHz and peaks around 25 kHz with 0.08 ng for the bundle with only tip links and 0.07 ng for the bundle with all elastic links. For higher frequencies they align. At 100 kHz the bundles without top connectors have an inertial mass of 0.048 ng and with top connectors 0.044 ng, values similar to the bundle mass.

5.2 Coupling between Stereocilia

Kozlov et al. (2007) used the coherence to evaluate the relative motion in a hair bundle based on experimentally measured time traces from the tall and short edge displacements in the stimulus direction of the hair bundle. This measure characterizes accurately the amount of common motion, even though the amplitudes might be scaled differently.[*]

The finite-element computations are performed in the frequency domain. Dr. Thomas Risler (Institut Curie, Physical Chemistry, Paris) provided the following relation to compute the coherency[†] from the model results (Kozlov et al. 2011).

The motions and forces in a given direction of two investigated stereocilia i and j are related to each other by the susceptibility matrix

$$\underbrace{\begin{bmatrix} \chi_{ii} & \chi_{ij} \\ \chi_{ji} & \chi_{jj} \end{bmatrix}}_{\chi} \begin{bmatrix} F_i \\ F_j \end{bmatrix} = \begin{bmatrix} u_i \\ u_j \end{bmatrix}. \quad (5.14)$$

All quantities are complex and frequency-dependent.

The susceptibility matrix is reconstructed from the model by two independent computations for each frequency. At first, a force is applied to the stereocilium i in the excitatory direction and the response is recorded in the same direction at both stereocilia i and j. In the second computation the excitation force is applied to the stereocilium j while again the displacements of both stereocilia i and j are recorded. This provides all needed information to reconstruct the susceptibility matrix χ in Equation (5.14).

---

[*]In the experimental situation it is difficult to record the time traces at the same height.

[†]The absolute value of the coherency is the coherence, which was used by Kozlov et al. (2007). In this work however the coherency is used to keep the sign information.

The spectral density matrix is given by the fluctuation-dissipation theorem based on the susceptibility matrix as

$$\mathbf{G} = 2 k_\mathrm{B} T \, \frac{\mathrm{Im}\,(\boldsymbol{\chi}(\omega))}{\omega} \qquad (5.15)$$

with the Boltzmann constant k_B, the temperature T, the circular frequency ω, and $\mathrm{Im}(\cdot)$ the imaginary part operator. Based on the spectral density matrix the coherency between the degrees of freedom i and j is given by

$$\gamma_{ij} = \frac{\mathbf{G}_{ij}}{\sqrt{\mathbf{G}_{ii}\,\mathbf{G}_{jj}}}. \qquad (5.16)$$

For a system operating at the thermodynamic equilibrium the imaginary part is always zero (De Groot and Mazur 1984). The real part is in the range of $-1 \leq \gamma \leq 1$.

Since the original matrices of stiffness, damping, and mass are symmetric, the susceptibility matrix should also be symmetric and $\gamma_{ij} = \gamma_{ji}$. A measure to control whether the symmetry is kept throughout the computation is the error

$$\varepsilon_\gamma = \mathrm{Abs}\left(2\,\frac{\gamma_{ji} - \gamma_{ij}}{\gamma_{ji} + \gamma_{ij}}\right). \qquad (5.17)$$

For the investigated frequency range and all different link combinations this ratio was always $\varepsilon_\gamma < 0.012$ and thus small enough to evaluate the coherency.

The probed stereocilia are both located on the symmetry plane of the bundle. The first one is the tallest stereocilium adjacent to the kinocilium. The other probed is the third shortest on the symmetry plane. The shortest stereocilium at the corner of the hexagon is frequently missing in real hair bundles. Computations based on the shortest stereocilium show a very specific behavior with low coupling. Taking the third shortest ensures more representative results for the coupling in the bundle. In this situation there are still three stereocilia standing between the two probed stereocilia.

In Figure 5.10 the displacement patterns for the two probed stereocilia are presented. For the stereocilium next to the kinocilium the results closely resemble the computations for an excitation at the kinociliary bulb (Figure 5.1) as the kinocilium and next neighbors are always strongly coupled by the kinociliary links. The excitation of the shortest stereocilium has a different displacement pattern for the whole bundle in the situation with pure fluid coupling (Figure 5.10a). At low frequencies the viscous coupling is insufficient to entrain the neighbors, but it becomes more efficient with increasing frequency. The bundle with only top connectors moves coherently for all frequencies and independent of the point of excitation. Only at very high frequencies of around 100 kHz at which inertial forces dominate and the displacement pattern depends on the point of excitation (Figure 5.10b). This is independent of the elastic linkage and so all displacement patterns with the same excitation location are the same for 100 kHz.

5.2 Coupling between Stereocilia

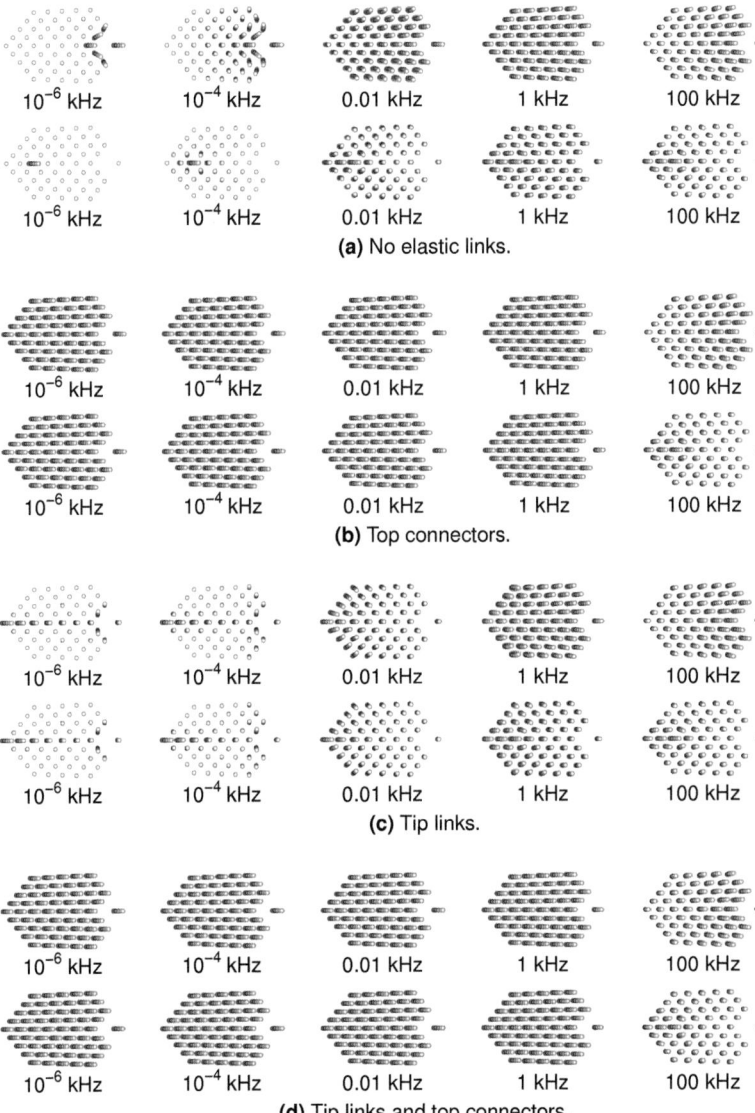

Figure 5.10: Time traces of the stereocilia for a displacement prescribed at stereocilia on the symmetry plane. The upper row is for the excitation of the next neighbor of the kinocilium. The lower row for the excitation of the third shortest. The displacements are scaled arbitrarily and divided by the individual stereocilia height for illustration purposes. The phase angle color code is given in Figure 5.1e. The kinocilium is located to the right. The time traces are centered around the location of the rootlets.

For intermediate frequencies the bundle with only tip links splays differently depending on the location of the excitation (Figure 5.10c). The bundle with all links present shows qualitatively the same motion patterns as the bundle with only top connectors (Figure 5.10d).

Just by visual interpretation it is hard to quantify the differences. The approach briefly outlined above from Dr. Thomas Risler (Institut Curie, Physical Chemistry, Paris) is used to evaluate the coherency and results are given in Figure 5.11. The bundle including all elastic links between stereocilia has a coherency of 0.94 for frequencies up to 0.1 kHz, which increases further to 0.97 at 1 kHz and falls off afterwards for frequencies higher than the undamped resonance frequency of the bundle. The evaluation of the drag was erroneous for frequencies below 0.1 Hz, if stiff top connectors were involved (Section 5.1). Therefore results were discarded for these low frequencies. If tip links are removed, the coherency increases for the frequencies below 1 kHz to 0.99, as no coupling elements induce relative motions between the stereocilia for a bundle deflection. If just tip links couple the stereocilia, the bundle remains coherent up to 1 Hz. Above this frequency viscous forces come into play and the stereocilia move with different amplitudes due to the oblique position of the tip links. The relative motion induced between stereocilia causes viscous forces with a different phase compared to the elastic coupling forces. The coherency decreases to 0.5 at 1 kHz and increases again slightly up to 0.63 at 6 kHz where viscous forces dominate. As for the other configurations, the coherency further decays above the undamped resonance frequency, where inertial forces come into play. The bundle without any elastic links has a coherency around zero up to 10 Hz. Above this frequency the coherency increases and is between 0.60 and 0.66 for frequencies between 0.1 and 10 kHz due to the viscous coupling forces between the stereocilia. These findings are in contrast to the results by Zetes and Steele (1997), who found a low-pass filter effect of the viscous liquid. Most probably this different behavior is due to the higher bundle stiffness and geometrical differences of the mammalian model they used.

The experimental data from Kozlov et al. (2007) show a high coherency of 0.95 ± 0.01 for an intact hair bundle undergoing Brownian motion and 0.97 ± 0.02 if all tip links are destroyed. Both results are in agreement with the model results. The high level of coherency for frequencies up to 10 kHz can only be reproduced with the numerical model if the top connectors are sufficiently stiff. Figure 5.12 illustrates the dependence on the top connector stiffness of a bundle without tip links. The top connectors start to influence the coherency with a stiffness of $20\ \mu\text{N}\cdot\text{m}^{-1}$ as their stiffness is high enough to overcome the overall bundle stiffness with the same value. The high coherency up to a few kilohertz is replicated if the coupling stiffness is further increased by a factor of a thousand to $20\ \text{mN}\cdot\text{m}^{-1}$.

5.2 Coupling between Stereocilia

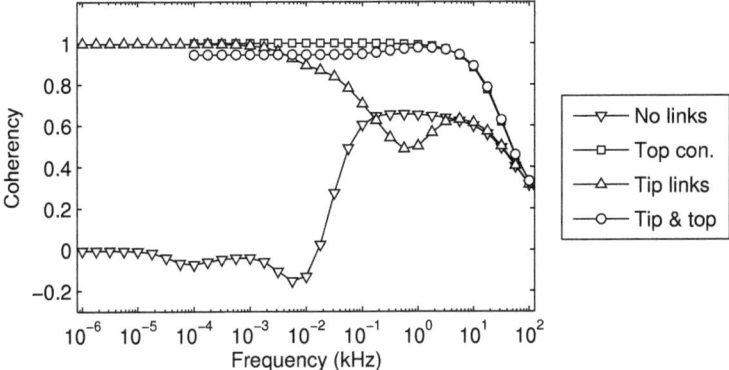

Figure 5.11: Coherency between the tallest and the third shortest stereocilium of the central column of the hair bundle. The results for the configurations including the top connectors are limited to a lower frequency of 0.1 Hz as the drag coefficient was erroneous for lower frequencies. Stereocilia are always coupled by the viscous liquid and elastically coupled only by horizontal top connectors (Top con.), tip links (Tip links), both types of elastic links (Tip & top), and no elastic links (No links).

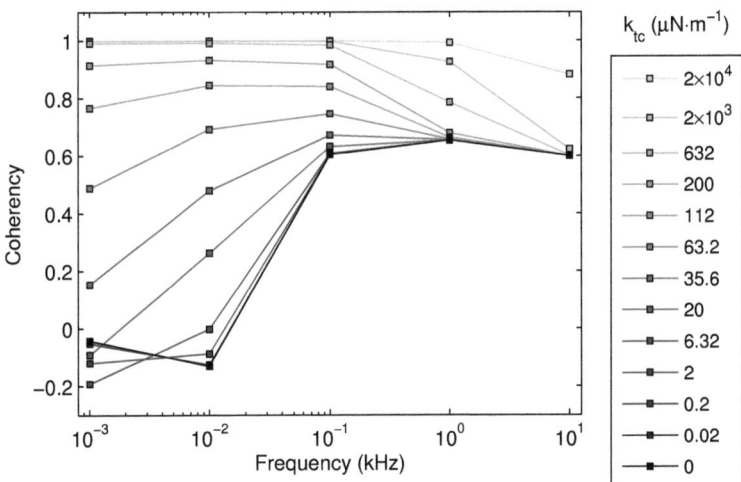

Figure 5.12: Coherency between the tallest and the third shortest stereocilium of the central column of the hair bundle. The tip links are removed and the top connector stiffness varies as given by the legend.

5.3 Response to Tip-Link Forces

Jaramillo et al. (1993) demonstrated that a nonlinearity inherent to the mechanoelectrical transduction process causes distortion products in hair cells. This is due to the gating of channels, which are mechanically tightly coupled to the tip links. This nonlinearity can be modeled by applying forces along the tip links as function of the bundle displacement. The finite-element model of the bundle cannot incorporate nonlinearities in its current state, but studying the linear response of the bundle to a force along the tip links helps to understand in which environment the channel gating forces work.

In vivo the bundle is attached to the otolithic membrane with a link of finite stiffness. Instead of a certain stiffness, two extreme situations are investigated here: the bundle is fully blocked in motion at the kinociliary bulb and the bundle is free to move. The first case corresponds to an infinite and the second to a zero coupling stiffness to a rigid and infinitely large otolithic membrane.

Figure 5.13 presents the bundle responses to the forces along all tip links with the same amplitude and phase to mimic a concerted gating of the tip links. For the boundary condition of a free moving bundle the kinocilial links were removed in all configurations. This ensured a free and undisturbed motion of the bundle. In the other situation the kinocilium was fixed in position and the kinocilial links were present to connect this constrain to the bundle at the tall edge. The excitation forces are applied at both ends of the links along the tip link axis in opposite directions.[*] The phase angle of zero degree corresponds to a maximum pulling force along the tip links.

If the bundle is solely coupled by fluid, it is displaced just at the outer edges (Figure 5.13a). With increasing frequency the motion enters gradually the bundle. For the held configuration only the next neighbors are affected by the presence of the kinocilium. With increasing frequency the orientations of motion of the stereocilia align with the direction away from the kinocilium. If the bundle is not constrained at the kinocilium, it splays away from a central point. This reference location shifts from the center towards the short edge with increasing frequency. The bundle with top connectors moves coherently in the unconstrained situation (Figure 5.13b). In the situation where the kinocilium is clamped, the bundle can splay with maximum amplitude only at the short edge. The displacement patterns at 100 kHz are an exception. At this high frequency the dominant inertial forces cause in all four link configurations similar displacement patterns. The bundle with tip links exhibits larger displacements at the short edge (Figure 5.13c), which is in agreement with the studies of the kinociliary displacement and coherence (Figures 5.1 and 5.10). The configuration with tip links and top

---

[*] In this study possible buckling of the tip links is not investigated. The tip links are assumed to be under high tension and the applied force along the tip link is regarded just as a small modulation of the always positive residual force.

5.3 Response to Tip-Link Forces

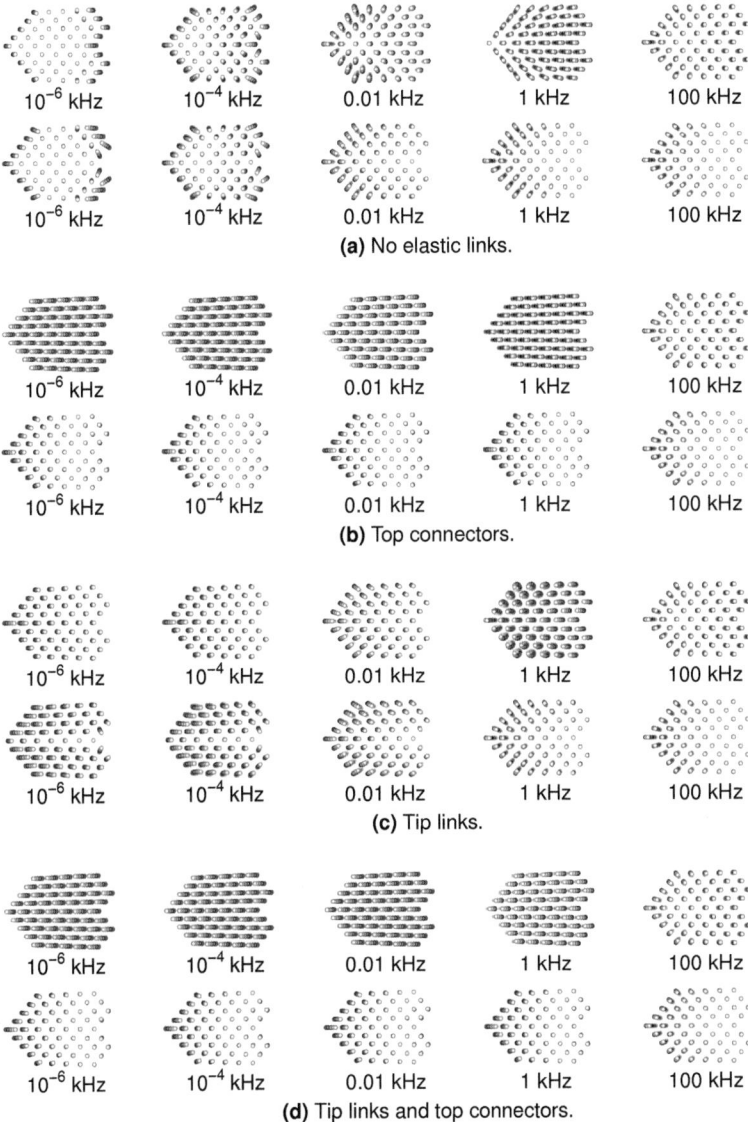

Figure 5.13: Time traces of the stereocilia for a force along all tip links with the same amplitude and phase. In each panel the upper row represents the case where the kinocilium is free to move. The lower row represents the case where the kinocilium is held in position at the kinociliary bulb. The kinocilium is not shown. The displacements are scaled arbitrarily and divided by the individual stereocilia height for illustration purposes. The phase angle color code is given in Figure 5.1e. The time traces are centered around the location of the rootlets.

connectors (Figure 5.13d) resembles the case with just top connectors and shows a coherent bulk motion for the free boundary condition. In the held configuration just the stereocilia around the short edge are moving.

The situation with the bundle held at the kinociliary bulb provides information on the relation between the force along the tip links and the force coupling to the otolithic membrane. In Figure 5.14 the reaction force at the held kinociliary bulb divided by the force applied along the tip links is plotted. Along all tip links a force with same amplitude and phase is applied. The reaction force is normalized by the force applied to a single tip link, and turns out to be around two.

The displacement patterns presented in Figure 5.13 show the relative motions. To quantify the displacements with respect to the applied force the susceptibility is presented in Figure 5.15 for the force applied along all tip links with the same phase and amplitude. The full bundle model contains 48 tip links. The susceptibility analyzed is the ratio of the displacement of the shortest stereocilium evaluated at the same height as the kinociliary bulb divided by the force applied to a single tip link. The ratio is complex. For all configurations the response is flat at low frequencies and decays with a power law of exponent around -1 at high frequencies. This indicates a stiffness dominated response for the low frequencies and a drag dominated for the higher frequencies. The main discrepancy between the different link combinations is the asymptotic value at the lowest frequencies. For the bundle just coupled by the fluid the susceptibility is 2.4 Mm·N^{-1}, independent of the boundary conditions at the kinocilium. If just tip links are present as elastic links, the value is 0.72 Mm·N^{-1} for the situation where the kinocilium is unconstrained and 0.14 Mm·N^{-1} if it is blocked. The bundle with only top connectors has a susceptibility of 0.22 Mm·N^{-1} in the free situation and a dramatically lower value of 0.88 km·N^{-1} with the held kinocilium. The later value is close to the susceptibility of the bundle including all links in the same situation, which yields 0.81 km·N^{-1}. If the kinocilium is not blocked, the susceptibility of the bundle with all links is 4.1 km·N^{-1}. Thus it is about 50 times easier to move the short edge of an intact bundle by forces along the tip links if the bundle is not held in position at the kinociliary bulb.

5.4 Analysis of Relative Motions in the Bundle based on Experimental Data

To validate the variation in susceptibility depending on the boundary condition at the kinociliary bulb valuable experimental data were provided by Dr. Andrei S. Kozlov (Laboratory of Sensory Neuroscience, Howard Hughes Medical Institute, Rockefeller

5.4 Analysis of Relative Motions in the Bundle

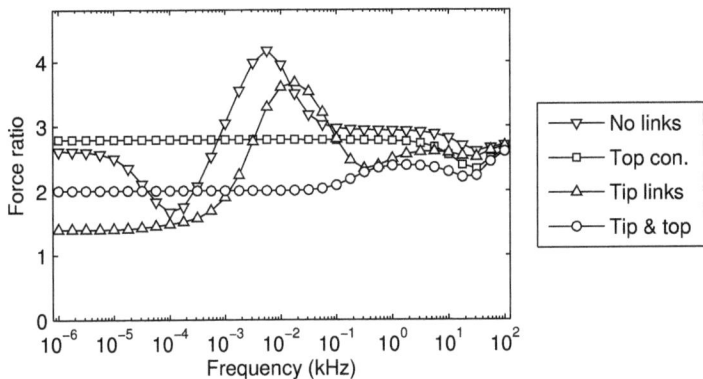

Figure 5.14: Ratio of the reaction force at the kinociliary bulb divided by the force applied along tip links. Stereocilia are always coupled by the viscous liquid and elastically coupled only by horizontal top connectors (Top con.), tip links (Tip links), both types of elastic links (Tip & top), and no elastic links (No links).

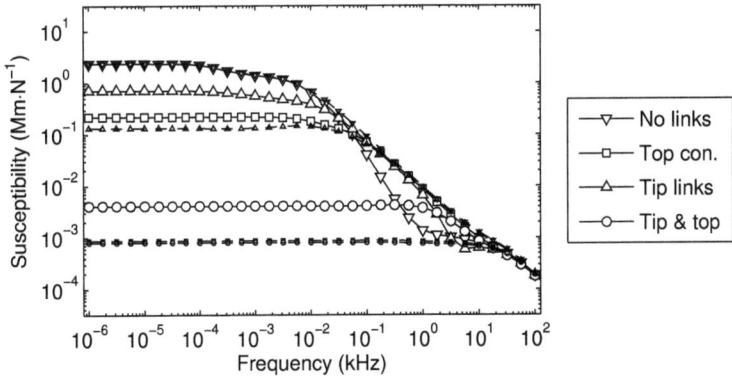

Figure 5.15: Susceptibility of the hair bundle at the short edge in response to forces applied along the location of all tip links with same phase and amplitude. The absolute value is given. The solid lines with large markers are for the bundle where the kinocilium is free to move. The dashed lines with small markers are for the situation where the kinocilium is held in position at the kinociliary bulb. Stereocilia are always coupled by the viscous liquid and elastically coupled only by horizontal top connectors (Top con.), tip links (Tip links), both types of elastic links (Tip & top), and no elastic links (No links).

University, New York). Together with him and Dr. Thomas Risler (Institut Curie, Physical Chemistry, Paris) a strategy was developed to analyze the experimental data and compare the results with the numerical model.

In the experiment a bundle was excited at the tall edge by two different types of probe attachment. In both cases a stiff probe was used. In the first condition the probe was firmly connected to the kinociliary bulb. In the second a loose coupling of the probe with the bundle was established. The first case prescribes the motion that the bundle has to follow. The second situation corresponds to an excitation by a force, such that motions different from the stimulus are still possible at the probe connection point. The analysis of the bundle response to these two different boundary conditions allows us to validate the numerical model of this work. The stereociliary coupling strength, mainly the top connecter stiffness, counteracts the nonlinearities introduced by the tip links. The value of this stiffness was set by adjusting to the experimental coherency measurements in Section 5.2.

To quantify the nonlinearities, the time series were analyzed separately and identically for the two edges of a hair bundle from which they were recorded simultaneously. The sampling of the data recording discretizes the time

$$t \to \tilde{\mathbf{t}} \tag{5.18}$$

into a vector with linearly increasing time from zero by steps of the reciprocal sampling frequency to the total measuring time. For a measured cell, twenty one-second-long measurements were aligned to the template $A_1 \sin(2\pi f_1 \tilde{\mathbf{t}}) + A_2 \sin(2\pi f_2 \tilde{\mathbf{t}})$, such that they all started at the same phase angle with respect to the motion at the tall edge. The two excitation frequencies are $f_1 = 90$ Hz and $f_2 = 115$ Hz, and the coefficients A_1 and A_2 are the particular amplitudes associated independently with each time trace. After the alignment, the traces were averaged to reduce the noise. The resulting sample was then subdivided in segments of 200 ms duration, which corresponded to the shortest time interval containing integer numbers of both stimulations' periods. After removing the onset and offset transients and subtracting the constant component of the signal in time, the average was taken over all subdivisions. This final mean signal $\tilde{\mathbf{u}}_m$ was then analyzed for its harmonic content. The amplitudes of the two primary frequencies were reconstructed by fitting to the undistorted reference displacement

$$\tilde{\mathbf{u}}_{\text{ref}} = B_1 \sin(2\pi f_1 \tilde{\mathbf{t}}) + B_2 \sin(2\pi f_2 \tilde{\mathbf{t}}) . \tag{5.19}$$

This signal $\tilde{\mathbf{u}}_{\text{ref}}$ then served as a reference to compute an orthonormal series of vectors $\mathbf{a}_i$ consisting of the power series of vectors $(\tilde{\mathbf{u}}_{\text{ref}})^i$, such that each $\mathbf{a}_i$ was a linear combination of the different $(\tilde{\mathbf{u}}_{\text{ref}})^j$, with $j \leq i$.* The orthogonality between the different $\mathbf{a}_i$ was ensured by the Gram-Schmidt algorithm. The projection of the averaged

---

*The notation $(\cdot)^i$ is used to indicate that the ith power has to be evaluated element-wise.

5.4 Analysis of Relative Motions in the Bundle

experimental displacement traces $\tilde{u}_m$ onto these basis vectors provided the expansion coefficients of the measured data as a power series of the undistorted input signal $\tilde{u}_{ref}$ as

$$\tilde{u}_m = C_1 \mathbf{a}_1 + C_2 \mathbf{a}_2 + \ldots = \sum_{i=0}^{\infty} C_i \mathbf{a}_i \quad \text{with} \quad \mathbf{a}_i^T \cdot \mathbf{a}_j = \delta_{ij} \quad (5.20)$$

where δ denotes the Kronecker delta function. The deviation from the reference signal is

$$\tilde{u}_{dis} = \tilde{u}_m - \tilde{u}_{ref} . \quad (5.21)$$

The coefficients $C_2, C_3, C_4, \ldots$ characterize the distortions. An example is given in

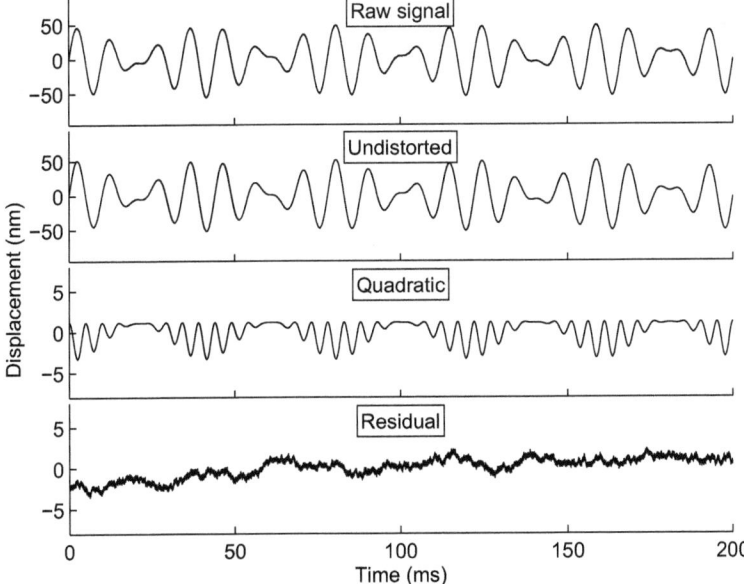

Figure 5.16: Time traces at short edge of a bundle stimulated with a firmly connected probe at the tall edge. The averaged original data from the experiment (Raw data) are analyzed by the fundamental signal (Undistorted), the quadratic distortions (Quadratic), and the remaining part of the original signal (Residual). *Experimental data courtesy of Dr. A. S. Kozlov, Rockefeller University*

Figure 5.16, where a time series of an experimental recording after averaging is presented. The linear contribution as well as the quadratic term and the residual are also plotted.

This procedure assumes that the distorted part of the measured signal $\tilde{u}_{dis}$ is solely a function of the reference signal $\tilde{u}_{ref}$ and thus the distortions are induced without any phase delay. A plot showing this distortion versus the reference signal $\tilde{u}_{ref}$ confirmers this hypothesis. In Figure 5.17 data and fits are shown for the firm and loose coupling of the probe at the short and tall edge.

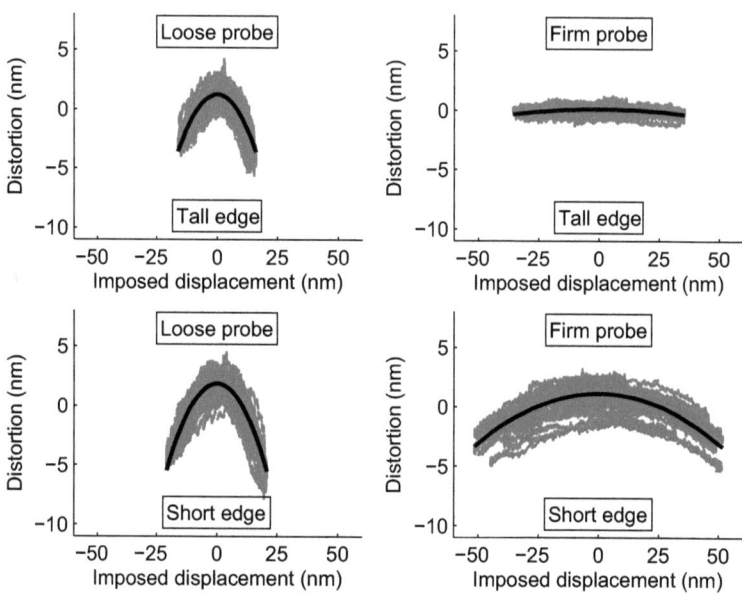

Figure 5.17: Distortion of the time traces at the tall and short edge of the bundle, while the bundle is excited with a probe by a loose or firm coupling. *Experimental data courtesy of Dr. A. S. Kozlov, Rockefeller University*

In the experiment with the firm probe the tall edge of a hair bundle was prevented from moving at the distortion-product frequencies. Therefore the focus is on the displacements at the opposite short edge. Although some cells demonstrated prominent quadratic and cubic distortions, in other cells, presumably because of a different operating point of the ion channels, the cubic distortion products were too small and buried in the noise. Therefore, only the first coefficient C_2 was significant for most of the experimental data, in that fitting to higher-order polynomials did not increase significantly the goodness of the fit.[*] Thus, the distortion $\tilde{u}_{dis}$ reduced under this approximation to a quadratic function of the linear part of the mean measured signal $\tilde{u}_m$ and in approxi-

---

[*]The fitting and comparison procedures were conducted with anova1 and multicompare functions in Matlab (Version 7.7, The MathWorks Inc., Natick, Massachusetts).

5.4 Analysis of Relative Motions in the Bundle

mation to a function of the reference signal $\tilde{u}_{ref}$ as

$$\tilde{u}_{dis} \approx \frac{(\tilde{u}_{ref})^2}{D_2} \,. \tag{5.22}$$

The coefficient D_2 has the unit of length and quantifies the distortion.

It is possible to reproduce the experimental observations with the finite-element model under the assumption that the distortion products originate from channel gating, and that this nonlinear response is a function of the bundle's deflection only. In Section 5.3 the bundle's response to a force exerted at both ends of a tip link in opposite directions along its length was discussed. The distortion stems from tip-link mechanics and is a quadratic function of the reference signal $\tilde{u}_{ref}$ (5.22). All other mechanical components of the system are linear here. With the ansatz that the tip-link force $\tilde{F}_{tl}$ is proportional to the distortion $\tilde{u}_{dis}$, and therefore quadratic in $\tilde{u}_{ref}$, the force that mimics the tip-links nonlinearity is

$$\tilde{F}_{tl} = \frac{(\tilde{u}_{ref})^2}{E_2} \,, \tag{5.23}$$

where E_2 is a scaling constant which has to be determined. The susceptibility for the motion at the edge in response to the force along the tip links is

$$\chi = \frac{\tilde{u}_{dis}}{\tilde{F}_{tl}} \,. \tag{5.24}$$

This provides a quadratic relation between the applied force along the tip links and the displacement at the short edge. Here, the susceptibility coefficient χ depends on both the tip-link and top-connector stiffness used in the model, as well as on the boundary conditions at the tall edge which is either loosely or firmly coupled to the probe. In the finite-element computation, these two experimental situations were mimicked by either leaving the tall edge free, or by blocking the motion there completely. Using Equations (5.21), (5.23), and (5.24) yields

$$E_2 = D_2 \, \chi \,. \tag{5.25}$$

Associated with the two experimental boundary conditions for which the two fitting coefficients D_2^{loose} and D_2^{firm} were obtained, the finite-element model provides the susceptibility coefficients χ^{loose} and χ^{firm}. As mentioned above, the tip-link force is assumed not to be a function of the boundary condition, but solely of the linear bundle displacement. Thus, the coefficient E_2 in Equation (5.23) is the same in both conditions, such that

$$E_2^{loose} = E_2^{firm} \,. \tag{5.26}$$

Using (5.25), the consequence is that the following ratio should be conserved between experimental and modeling data as

$$\frac{D_2^{\text{loose}}}{D_2^{\text{firm}}} = \frac{\chi^{\text{firm}}}{\chi^{\text{loose}}}. \qquad (5.27)$$

For a given reference signal $\tilde{\mathbf{u}}_{\text{ref}}$, when the tall edge is free to move at the distortion frequencies, tip links have to work mostly against the small inter-stereocilia sliding drag and the tip links' stiffness to create a given distortion signal $\tilde{\mathbf{u}}_{\text{dis}}$ at the hair bundle's short edge; but if the tall edge is constrained, the same tip links have to work in addition against the larger inter-stereocilia squeezing drag and against the top connectors' stiffness. Therefore, the corresponding $\tilde{\mathbf{u}}_{\text{dis}}$ is smaller in the second case, which according to Equation (5.22) corresponds to a larger D_2. Consequently, the ratio given above is smaller than one.

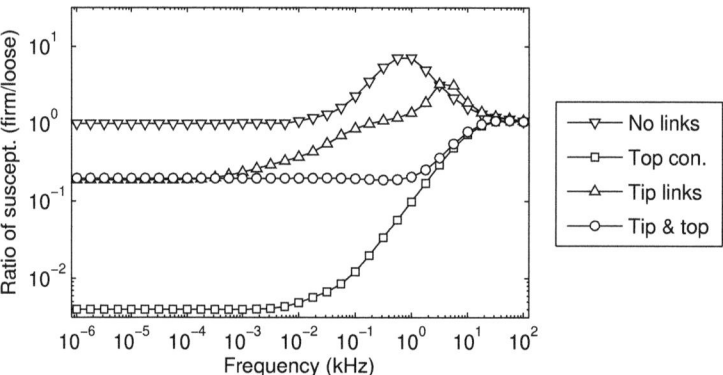

Figure 5.18: Ratio of the susceptibilities at the short edge. The ratio is the susceptibility of the bundle with the kinocilium held divided by the bundle with the kinocilium free to move. Stereocilia are always coupled by the viscous liquid and elastically coupled only by horizontal top connectors (Top con.), tip links (Tip links), both types of elastic links (Tip & top), and no elastic links (No links).

Figure 5.18 shows the ratio of the susceptibilities from the finite-element model. The data for the susceptibilities are from Figure 5.15. The purely fluid-coupled bundle has a value around 1 for most frequencies. For the bundle with top connectors the ratio is 0.004 for the low frequencies, starting from 0.1 kHz the ratio increases to about 1. The bundle with tip links has a ratio of 0.19 at low frequencies starting from 1 Hz and increases up to a peak of about 3 around 4 kHz. The intact bundle with all links has a value of 0.20 for frequencies up to 1 kHz. At higher frequencies the ratio aligns with the ratio of the other configurations. The intact bundle configuration is to be compared

5.4 Analysis of Relative Motions in the Bundle

with the experimental data for frequencies around the experimental frequencies that correspond to quadratic distortions in the range from 25 Hz to 230 Hz.

The analysis of the experimental data yields $D_2^{loose} = 0.14 \pm 0.12$ µm ($n = 8$) for the loose attachment of the probe to the bundle and $D_2^{firm} = 1.6 \pm 0.9$ µm ($n = 4$) for the firm coupling to the bundle (Kozlov et al. 2011). The ratio of these two mean values is 0.09 with a high uncertainty, but close to the ratio of the model with a value of 0.20. This validates the model further and shows that the stiff top connectors in the bundle do not totally prevent relative motions in the hair bundle. The differences in the distortions between the two bundle edges for the experiments with a firm probe measures about a twentieth of the amplitude of 50 nm of the undistorted signal for the relative motion (Figure 5.17). This is a remarkably low relative motion for a single pair of stereocilia, as this has to be divided by the number of pairs between the edges. This yields an experimentally measured relative motion in the order of 0.1 nm, which is at the scale of the Bohr radius.

Chapter 6

Conclusion

6.1 Summary

In this work the fluid and solid mechanics of the hair bundle from the bullfrog's sacculus are numerically analyzed in detail. The developed physical model encompasses all relevant scales from the tiny gaps between the stereocilia to the viscous boundary layer around the bundle. The geometry of the model captures uncompromisingly the dimensions of the bundle *in vivo* with the individual stereocilia, the kinocilium, and the connecting links. The elastic solid material is formulated as isotropic and in terms of displacement as variable. A finite-element formulation for the liquid with displacement and pressure variables permits computation of the strongly coupled fluid-structure interaction problem with a single solution of the system matrix for a wide range of material properties. The implementation is successfully validated with test cases encountering inertial, viscous, and elastic forces. The unified formulation for fluid and solid is introduced by discussing the governing equations. The use of appropriate dimensionless numbers and order of magnitude estimates allows to simplify and linearize the system of equations.

The hair bundle is analyzed in detail under different load conditions. Stereocilia in a hair bundle are connected by oblique tip links and by the horizontal top connectors, the stiffness of which has not been measured directly. The only way to estimate their properties is by well-parameterized models. So far no other model exists incorporating the full fluid-structure interaction of the hair-bundle. The agreement with various different experimental data confirms a correct representation of the mechanics in the numerical model and therefore a good estimation of the properties of the underlying structure. A summary of the main results from each chapter follows in the order of their appearance.

In Chapter 2, the equations of fluid mechanics are analyzed with respect to the specific physical scales of hearing mechanics and the small amplitudes of motion. Order-of-magnitude estimates allow us to neglect the convective terms and the energy equation. The velocity can be substituted by the time derivative of the displacement.

The solid material is also described by the displacement, a common practice in solid mechanics. It was advantageous for the numerical computation to keep the pressure-displacement relation for the fluid separated. Both field equations for the displacement of fluid and solid rely on the conservation of momentum and differ only in the material law.

In Chapter 3, the numerical realization of the differential field equations of Chapter 2 is demonstrated by means of the finite-element method. The fluid equations are discretized with quadratic basis functions and the pressure with linear basis functions. The implementation is validated by three representative test cases. Firstly, the force needed to drive an oscillating sphere is used to check the viscous and inertial forces. Secondly, the impedance of a closed tube provides a validation of the compressive and inertial forces, influenced slightly by the viscous boundary layer. Thirdly, the volumetric compression of an elastic shell filled with fluid serves as a control on the momentum and displacement transfer between solid and fluid. In all three cases the validation has been successful. Therefore this approach is not only suitable for problems in hearing mechanics, but also for all mechanical situations in which small-amplitude oscillatory motions are at play. Examples are the air oscillations in musical instruments dampened by a thin viscous boundary layer, atomic-force microscopes in liquid environments where the resonance frequencies are highly affected by the added-mass effect and damped as well, and the absorption of sound in elastic structures with small pores in which the pressure field displaces the structure and dissipation takes place in the fluid.

In Chapter 4, a finite-element model of the hair bundle of the bullfrog's sacculus is constructed. Geometrical data are taken from literature and further detailed information is obtained from micrographs. The material properties are selected to represent various different experimental observations. The pivotal stiffness of the stereocilia is set to 10 aN·m·rad^{-1}, the flexural rigidity of the kinocilium to 630 yN·m^2, the top-connector stiffness to 20 mN·m^{-1}, and the tip-link stiffness to 1 mN·m^{-1}. The geometry is meshed with the finite-element formulation presented in Chapter 3 for the liquid. For the isotropic solid and the coupling links standard linear elastic elements are used. The finite-element mesh so generated is multi-block structured and uses hexahedral elements only. The mesh is parametric and can be refined as needed.

In Chapter 5, a detailed analysis is conducted for the mechanical forces necessary to deflect the bundle, for the mechanical coupling in the bundle, and for the response of the bundle to a tip-link force. The drag of the bundle without tip links is evaluated to be about 55 nN·s·m^{-1} by removal of the tip links in the model. The addition of tip links causes small additional relative motions associated with a high squeezing drag in between stereocilia and the drag increases to about 85 nN·s·m^{-1}. The drag induced by the external liquid is around 30 nN·s·m^{-1} including the sliding drag of about 10 nN·s·m^{-1} inside the bundle. The drag of an individual stereocilium is computed to be about

10 nN·s·m^{-1}. This result reveals that grouping stereocilia in a bundle dramatically reduces the total drag as compared to the sum of about 60 individual stereocilia moving in isolation in the liquid. The stiffness of the bundle is about 20 µN·m^{-1} without and 440 µN·m^{-1} with the tip links. The bundle with only liquid in between the stereocilia displays a coherence of about 0.6 for frequencies above 0.1 kHz. A high coherence of about 0.99 is achieved only with a high top-connector stiffness of 20 mN·m^{-1} for frequencies from 10 Hz to 10 kHz and is almost fully retained with a coherence of about 0.95 if the tip links are added to the bundle. The analysis of experiments with distortions generated at the tip links by the gating compliance confirmed the observed reduction in relative motion inside the bundle, when the probe attachment is changed from loose to firm.

In summary, this work presents what is currently the most detailed, realistic linear model of hair-bundle mechanics, that explicitly takes into account the fluid-structure interactions between individual stereocilia. This reveals how the basic structural features of a hair bundle interact in the process of hearing.

6.2 Directions for Future Research

This work provides new insights into the mechanics of the hair bundle and a novel approach to model the fluid-structure interaction for hearing mechanics and similar mechanical problems. Not all questions of hair-bundle mechanics could be addressed and the method presented for the analysis of fluid-structure interaction could be further improved, for example in its numerical implementation, or by using a more general material description, and in the incorporation of non-linearities.

The numerical method is currently restricted to linear mechanical situations. It might be of interest to extend the range to small non-linearities by a fixed-point expansion to allow larger displacements with respect to the geometry. This would provide a basis to investigate steady flow patterns generated by oscillatory motions. For objects surrounded by an infinite amount of fluid, it would be advantageous to use non-reflecting boundary conditions to decrease the size of the problem to solve. So far, a direct solver was used to solve the system of equations of the finite element problem. An appropriate pre-conditioner and a suitable iterative solver might significantly accelerate the solution procedure.

Care was taken to reconstruct the geometry as realistically as possible. Some assumptions were still necessary; they can be verified when new imaging techniques become available. Additional information about the space between individual stereocilia would be especially helpful. Although the structure of the bullfrog's saccular hair bundle has similarities with that of the mammalian hair bundle, there are also differences. A model of the mammalian outer and inner hair cell bundles would provide the possibility to further investigate their mechanics. For the inner hair cell bundle the displacement in a fluid flow would be of interest.

The material properties of the stereocilia are assumed so far to be isotropic, but the ultrastructure shows periodic structures with a distinct orientation. Based on mechanical measurements of the stereocilia it might be possible to identify the orthotropic elastic and damping properties for the material and thus to provide the information necessary to further adjust the model to the *in vivo* hair bundle. Also, further experimental characterization of the links and other coupling effects between stereocilia would be important for future models of the hair bundle. The no-slip condition at the interface of solid and liquid needs further clarification. The properties of the endolymph are assumed as viscous, but at low shear rates the liquid might be viscoelastic. Here further experimental data are needed.

The finite-element model is very detailed and provides a large amount of information. For studies of the bundle interacting with the environment it will be necessary to cast the dynamical behavior of the bundle into a model with just a few degrees of freedom. The reduction of the problem size will also provide a basis to model channel gating in a realistic mechanical environment and the response of the motion of the bundle in a thermal bath.

6.3 Acknowledgments

I would like to express my sincere thanks and gratitude to my friends and colleagues who have followed and supported me throughout this work.* Without them this work would simply not have been possible. Here I can just briefly mention a few.

It is a pleasure to thank those who made this work possible. I owe my deepest gratitude to Prof. Roger Grundmann for giving me the opportunity to work with freedom on hearing and musical acoustics, to Prof. A. J. Hudspeth for confidence in my work and support to conduct research in the fascinating world of the hair bundle, to Prof. Axel Voigt for providing an inspiring and supportive environment to work on this interdisciplinary subject, to Dr. Andrei S. Kozlov for comprehensive support on the detailed aspects of the hair bundle and always interpreting the mysterious results, to Dr. Thomas Risler for the theoretical background of the stochastic problems and taking care about all the theoretical equations, to Mario Fleischer for discussing solid and fluid mechanics and the finite-element implementation deeply in the context of hearing mechanics, and to Roland Gärtner for the work and discussion on the model of the organ of Corti and on the material of the stereocilia. Furthermore I am grateful for support by: Catrin Mühlfriedel, Timo Grothe, and Dr. Andreas Richter. For helpful discussions on the theory, experiments in the context of the real hair bundle I would like to

---

*This work was also supported by DFG Gr 1388/14, by DAAD D/07/46200, and by the resources of the Center for Information Services and High Performance Computing of the Technische Universität Dresden.

6.3 Acknowledgments

show my gratitude to: Dr. Daniel Andor, Adria LeBœuf, Alexander Darbinjan, Dr. Kai Dierkes, Prof. Jonathan Howard, Prof. Frank Jülicher, Prof. Holger Kantz, Dr. Benjamin Lindner, Dr. Dáibhid Ó Maoiléidigh, Dr. Pascal Martin, Prof. Sunil Puria, and Prof. Charles R. Steele.

The basis for the work was provided by an inspiring collaborative research project on the mechanics of the inner ear. I would like to express my thanks to: Matthias Bornitz, Caio Chiaradia, Daniel Gloger, Prof. Anthony W. Gummer, Dr. Csaba Harasztosi, Prof. Hans-Jürgen Hardtke, Claudia Hertel, Prof. Gert Hofmann, Dr. Manuela Nowotny, Anton A. Poznyakovskiy, Prof. Rolf Schmidt, Dr. Hannes Seidler, Corstiaen P. C. Versteegh, Dr. Aleš Vetešník, Dr. Yury M. Yarin, and Prof. Thomas Zahnert.

My thanks to all of my friends, my family, and especially Ruth.

Bibliography

J. D. Anderson, G. Degrez, J. Degroote, E. Dick, R. Grundmann, and J. Vierendeels. *Computational Fluid Dynamics: An Introduction (Von Karman Institute Book)*. Springer, Berlin, 3rd edition, 2009.

M. Andoh and H. Wada. Prediction of the characteristics of two types of pressure waves in the cochlea: Theoretical considerations. *The Journal of the Acoustical Society of America*, 116(1):417–425, 2004.

M. Andoh, C. Nakajima, and H. Wada. Phase of neural excitation relative to basilar membrane motion in the organ of Corti: Theoretical considerations. *The Journal of the Acoustical Society of America*, 118(3):1554–1565, 2005.

J. Ashmore, P. Avan, W. Brownell, P. Dallos, K. Dierkes, R. Fettiplace, K. Grosh, C. Hackney, A. J. Hudspeth, F. Jülicher, B. Lindner, P. Martin, J. Meaud, C. Petit, J. R. Santos Sacchi, and B. Canlon. The remarkable cochlear amplifier. *Hearing Research*, 266(1–2):1–17, 2010.

M. E. Bashtanov, R. J. Goodyear, G. P. Richardson, and I. J. Russell. The mechanical properties of chick (Gallus domesticus) sensory hair bundles: relative contributions of structures sensitive to calcium chelation and subtilisin treatment. *The Journal of Physiology*, 559(Pt 1):287–299, 2004.

M. Bathe, C. Heussinger, M. M. A. E. Claessens, A. R. Bausch, and E. Frey. Cytoskeletal bundle mechanics. *Biophysical Journal*, 94(8):2955–2964, 2008.

J. Baumgart, M. Fleischer, Y. M. Yarin, and R. Grundmann. Fluid flow around the stereocilia. In X. Ottavy and I. Trebinjac, editors, *8th International Symposium on Experimental and Computational Aerothermodynamics of Internal Flows*, volume 1, pages 189–193, 2007.

J. Baumgart, C. Chiaradia, M. Fleischer, Y. M. Yarin, R. Grundmann, and A. W. Gummer. *Concepts and Challenges in the Biophysics of Hearing*, chapter Fluid mechanics in the subtectorial space, pages 288–293. World Scientific Press, New Jersey, London, Singapore, Beijing, Shanghai, Hong Kong, Taipei, Chennai, 1st edition, 2009a.

J. Baumgart, A. S. Kozlov, T. Risler, R. Gärtner, C. P. C. Versteegh, M. Fleischer, A. Voigt, and A. J. Hudspeth. Fluid forces in the hair bundle of the bullfrog's sacculus. In *International Conference on Natural and Biomimetic Mechanosensing*, 2009b.

J. Baumgart, M. Fleischer, and A. Voigt. Fluid-structure interaction in the hair bundle modelled by the finite-element method. In *US National Committee on Theoretical and Applied Mechanics*, 2010.

S. Broersma. Rotational diffusion constant of a cylindrical particle. *The Journal of Chemical Physics*, 32(6):1626–1631, 1960.

S. Broersma. Viscous force and torque constants for a cylinder. *The Journal of Chemical Physics*, 74(12):6989–6990, 1981.

H. Cai, B. Shoelson, and R. S. Chadwick. Evidence of tectorial membrane radial motion in a propagating mode of a complex cochlear model. *Proceedings of the National Academy of Sciences of the United States of America*, 101(16):6243–6248, 2004.

C. Chiaradia, M. Nowotny, and A. W. Gummer. *Concepts and Challenges in the Biophysics of Hearing*, chapter Deflection of IHC stereocilia in response to somatic OHC electromotility, pages 283–287. World Scientific Press, New Jersey, London, Singapore, Beijing, Shanghai, Hong Kong, Taipei, Chennai, 1st edition, 2009.

J. Cotton and W. Grant. Computational models of hair cell bundle mechanics: II. Simplified bundle models. *Hearing Research*, 197(1-2):105–111, 2004.

A. C. Crawford and R. Fettiplace. The mechanical properties of ciliary bundles of turtle cochlear hair cells. *The Journal of Physiology*, 364:359–379, 1985.

D. Dahl and D. Kleinfeldt. Method for measuring the viscosity of the perilymph in the guinea pig cochlea under physiological conditions. *Archiv für klinische und experimentelle Ohren- Nasen- und Kehlkopfheilkunde*, 197(1):31–40, 1970.

S. R. De Groot and P. Mazur. *Non-Equilibrium Thermodynamics*. Dover Publications, Mineola, 1st edition, 1984.

W. Denk, W. W. Webb, and A. J. Hudspeth. Mechanical properties of sensory hair bundles are reflected in their brownian motion measured with a laser differential interferometer. *Proceedings of the National Academy of Sciences of the United States of America*, 86(14):5371–5375, 1989.

K. Dierkes, B. Lindner, and F. Jülicher. Enhancement of sensitivity gain and frequency tuning by coupling of active hair bundles. *Proceedings of the National Academy of Sciences of the United States of America*, 105(48):18669–18674, 2008.

M. Fleischer, R. Schmidt, and A. W. Gummer. Compliance profiles derived from a three-dimensional finite-element model of the basilar membrane. *The Journal of the Acoustical Society of America*, 127(5):2973–2991, 2010.

D. M. Freeman and T. F. Weiss. Hydrodynamic analysis of a two-dimensional model for micromechanical resonance of free-standing hair bundles. *Hearing Research*, 48 (1-2):37–67, 1990.

J. A. García, A. G. Yee, P. G. Gillespie, and D. P. Corey. Localization of myosin-Iβ near both ends of tip links in frog saccular hair cells. *The Journal of Neuroscience*, 18(21):8637–8647, 1998.

D. R. Gaskell. *Introduction to the Thermodynamics of Materials*. Taylor & Francis, New York, London, 5th edition, 2008.

C. D. Geisler. A model of stereociliary tip-link stretches. *Hearing Research*, 65(1-2): 79–82, 1993.

P. G. Gillespie and U. Müller. Mechanotransduction by hair cells: models, molecules, and mechanisms. *Cell*, 139(1):33–44, 2009.

F. Gittes, B. Mickey, J. Nettleton, and J. Howard. Flexural rigidity of microtubules and actin filaments measured from thermal fluctuations in shape. *The Journal of Cell Biology*, 120(4):923–934, 1993.

T. Gold. Hearing. II. The physical basis of the action of the cochlea. *Proceedings of the Royal Society of London. Series B, Biological Sciences*, 135:492–498, 1948.

R. E. Graves and B. M. Argrow. Bulk viscosity: Past to present. *Journal of Thermophysics and Heat Transfer*, 13(3):337–342, 1999.

C. J. Greenshields and H. G. Weller. A unified formulation for continuum mechanics applied to fluid-structure interaction in flexible tubes. *International Journal for Numerical Methods in Engineering*, 64(12):1575–1593, 2005.

J. Howard. *Mechanics of Motor Proteins and the Cytoskeleton*. Sinauer Associates, Sunderland, Massachusetts, 1st edition, 2001.

J. Howard and J. F. Ashmore. Stiffness of sensory hair bundles in the sacculus of the frog. *Hearing Research*, 23(1):93–104, 1986.

J. Howard and A. J. Hudspeth. Compliance of the hair bundle associated with gating of mechanoelectrical transduction channels in the bullfrog's saccular hair cell. *Neuron*, 1(3):189–199, 1988.

J. Howard, W. M. Roberts, and A. J. Hudspeth. Mechanoelectrical transduction by hair cells. *Annual Review of Biophysics and Biophysical Chemistry*, 17:99–124, 1988.

A. J. Hudspeth. How the ear's works work. *Nature*, 341(6241):397–404, 1989.

A. J. Hudspeth. Making an effort to listen: mechanical amplification in the ear. *Neuron*, 59(4):530–545, 2008.

A. J. Hudspeth, F. Jülicher, and P. Martin. A critique of the critical cochlea: Hopf–a bifurcation–is better than none. *Journal of Neurophysiology*, 104(3):1219–1229, 2010.

A. J. Hunt, F. Gittes, and J. Howard. The force exerted by a single kinesin molecule against a viscous load. *Biophysical Journal*, 67(2):766–781, 1994.

R. A. Jacobs and A. J. Hudspeth. Ultrastructural correlates of mechanoelectrical transduction in hair cells of the bullfrog's internal ear. *Cold Spring Harbor Symposia on Quantitative Biology*, 55:547–561, 1990.

A. F. Jahn and J. Santos-Sacchi. *Physiology of the Ear*. Singular, San Diego, 2nd edition, 2001.

F. Jaramillo and A. J. Hudspeth. Displacement-clamp measurement of the forces exerted by gating springs in the hair bundle. *Proceedings of the National Academy of Sciences of the United States of America*, 90(4):1330–1334, 1993.

F. Jaramillo, V. S. Markin, and A. J. Hudspeth. Auditory illusions and the single hair cell. *Nature*, 364(6437):527–529, 1993.

D. J. Jeffrey and Y. Onishi. The slow motion of a cylinder next to a plane wall. *The Quarterly Journal of Mechanics and Applied Mathematics*, 34(2):129–137, 1981.

J. Kestin and W. A. Wakeham. *Transport Properties of Fluids: Thermal Conductivity, Viscosity, and Diffusion Coefficient*. Center for Information and Numerical Data Analysis and Synthesis. Hemisphere, New York, 1st edition, 1988.

G. H. Keulegan and L. H. Carpenter. Forces on cylinders and plates in an oscillating fluid. *Journal of Research of the National Bureau of Standards*, 60(5):423–440, 1958.

G. R. Kirchhoff. Ueber den Einfluss der Wärmeleitung in einem Gase auf die Schallbewegung. *Annalen der Physik und Chemie*, 2(134):177–193, 1868.

A. S. Kozlov, T. Risler, and A. J. Hudspeth. Coherent motion of stereocilia assures the concerted gating of hair-cell transduction channels. *Nature Neuroscience*, 10(1):87–92, 2007.

A. S. Kozlov, J. Baumgart, T. Risler, C. P. C. Versteegh, and A. J. Hudspeth. Fluid-mediated coupling and drag reduction in arrays of closely apposed stereocilia. *The Journal of Physiological Sciences*, 59(Supplement 1):201, 2009.

A. S. Kozlov, J. Baumgart, T. Risler, C. P. C. Versteegh, and A. J. Hudspeth. Forces between clustered stereocilia minimize friction in the ear on a subnanometre scale. *Nature*, 474(7351):376–379, 2011.

E. Lauga, M. P. Brenner, and H. A. Stone. *Handbook of Experimental Fluid Dynamics*, chapter Microfluidics: The no-slip boundary condition, pages 1219–1240. Springer, New York, 1st edition, 2005.

E. W. Lemmon, M. O. McLinden, and D. G. Friend. *NIST Chemistry WebBook, NIST Standard Reference Database Number 69*, chapter Thermophysical Properties of Fluid Systems. National Institute of Standards and Technology, Gaithersburg, 2010. URL http://webbook.nist.gov. retrieved June 2, 2010.

D. R. Lide. *CRC Handbook of Chemistry and Physics*. Taylor & Francis, 90th edition, 2009.

P. Martin, A. D. Mehta, and A. J. Hudspeth. Negative hair-bundle stiffness betrays a mechanism for mechanical amplification by the hair cell. *Proceedings of the National Academy of Sciences of the United States of America*, 97(22):12026–12031, 2000.

P. Martin, A. J. Hudspeth, and F. Jülicher. Comparison of a hair bundle's spontaneous oscillations with its response to mechanical stimulation reveals the underlying active process. *Proceedings of the National Academy of Sciences of the United States of America*, 98(25):14380–14385, 2001.

T. Matsui, C. Nakajima, Y. Yamamoto, M. Andoh, K. Iida, M. Murakoshi, S. Kumano, and H. Wada. Analysis of the dynamic behavior of the inner hair cell stereocilia by the finite element method. *JSME international Journal*, 49(3):828–836, 2006.

B. Nadrowski, P. Martin, and F. Jülicher. Active hair-bundle motility harnesses noise to operate near an optimum of mechanosensitivity. *Proceedings of the National Academy of Sciences of the United States of America*, 101(33):12195–12200, 2004.

J.-H. Nam, J. R. Cotton, and J. W. Grant. Effect of fluid forcing on vestibular hair bundles. *Journal of Vestibular Research*, 15:263–278, 2005.

J.-H. Nam, J. R. Cotton, E. H. Peterson, and W. Grant. Mechanical properties and consequences of stereocilia and extracellular links in vestibular hair bundles. *Biophysical Journal*, 90(8):2786–2795, 2006.

J.-H. Nam, J. R. Cotton, and W. Grant. A virtual hair cell, I: addition of gating spring theory into a 3-D bundle mechanical model. *Biophysical Journal*, 92(6):1918–1928, 2007.

C. J. Nederveen. *Acoustical Aspects of Woodwind Instruments*. Northern Illinois University Press, reprinted edition, 1998.

I. Newton. *Philosophiæ naturalis principia mathematica*, chapter Philiosophiae naturalis de motu corporum. Liber secundus. Sectio IX. De motu circulari fluidorum, pages 374–385. Innys, 1st edition, 1726.

A. V. Nguyen. Historical note on the Stefan-Reynolds equations. *Journal of Colloid and Interface Science*, 231(1):195–195, 2000.

M. Nowotny and A. W. Gummer. Nanomechanics of the subtectorial space caused by electromechanics of cochlear outer hair cells. *Proceedings of the National Academy of Sciences of the United States of America*, 103(7):2120–2125, 2006.

M. Okuno and Y. Hiramoto. Direct measurements of the stiffness of echinoderm sperm flagella. *Journal of Experimental Biology*, 79:235–243, 1979.

G. Papadakis. A novel pressure-velocity formulation and solution method for fluid-structure-interaction problems. *Journal of Computational Physics*, 227(6):3383–3404, 2008.

F. Perrin. Mouvement brownien d'un ellipsoide - I. Dispersion diélectrique pour des molécules ellipsoidales. *Le Journal de Physique et le Radium*, 5(10):497–511, 1934.

J. O. Pickles. A model for the mechanics of the stereociliar bundle on acousticolateral hair cells. *Hearing Research*, 68(2):159–172, 1993.

A. D. Pierce. *Acoustics: An Introduction to Its Physical Principles and Applications*. American Institute of Physics, reprinted edition, 1994.

A. A. Poznyakovskiy, T. Zahnert, Y. Kalaidzidis, R. Schmidt, B. Fischer, J. Baumgart, and Y. M. Yarin. The creation of geometric three-dimensional models of the inner ear based on micro computer tomography data. *Hearing Research*, 243(1-2):95–104, 2008.

C. Pozrikidis. *Introduction to Finite and Spectral Element Methods using Matlab*. Chapman and Hall/CRC, 1st edition, 2005.

S. Puria and C. R. Steele. *Handbook of the Senses: Audition*, chapter Mechano-Acoustical Transformations, pages 165–201. Elsevier, 1st edition, 2008.

S. Ramamoorthy, N. V. Deo, and K. Grosh. A mechano-electro-acoustical model for the cochlea: response to acoustic stimuli. *The Journal of the Acoustical Society of America*, 121(5):2758–2773, 2007.

A. J. Ricci, A. C. Crawford, and R. Fettiplace. Active hair bundle motion linked to fast transducer adaptation in auditory hair cells. *The Journal of Neuroscience*, 20(19):7131–7142, 2000.

L. Robles and M. A. Ruggero. Mechanics of the mammalian cochlea. *Physiological Reviews*, 81(3):1305–1352, 2001.

H. Schlichting and K. Gersten. *Boundary-Layer Theory*. Springer, Berlin, Heidelberg, 8th edition, 2000.

L. F. Shatz. The effect of hair bundle shape on hair bundle hydrodynamics of nonmammalian inner ear hair cells for the full frequency range. *Hearing Research*, 195 (1-2):41–53, 2004.

J.-B. Shin, F. Streijger, A. Beynon, T. Peters, L. Gadzala, D. McMillen, C. Bystrom, C. E. E. M. Van der Zee, T. Wallimann, and P. G. Gillespie. Hair bundles are specialized for ATP delivery via creatine kinase. *Neuron*, 53(3):371–386, 2007.

SI. The international system of units (SI). Technical report, Bureau International des Poids et Mesures, 2006.

J. Silber, J. Cotton, J.-H. Nam, E. H. Peterson, and W. Grant. Computational models of hair cell bundle mechanics: III. 3-D utricular bundles. *Hearing Research*, 197(1-2): 112–130, 2004.

M. Sotomayor, W. A. Weihofen, R. Gaudet, and D. P. Corey. Structural determinants of cadherin-23 function in hearing and deafness. *Neuron*, 66(1):85–100, 2010.

C. R. Steele, J. Boutet de Monvel, and S. Puria. A multiscale model of the organ of Corti. *Journal of Mechanics of Materials and Structures*, 4(4):755–778, 2009.

G. G. Stokes. On the effect of the internal friction of fluids in motion and of the equilibrium and motion of elastic solids. *Transactions of the Cambridge Philosophical Society*, 8:287–319, 1849.

G. G. Stokes. On the effect of the internal friction of fluids on the motion of pendulums. *Transactions of the Cambridge Philosophical Society*, 9(2):8–106, 1856.

M. Thorne, A. N. Salt, J. E. DeMott, M. M. Henson, O. W. J. Henson, and S. L. Gewalt. Cochlear fluid space dimensions for six species derived from reconstructions of three-dimensional magnetic resonance images. *The Laryngoscope*, 109(10):1661–1668, 1999.

J. A. Tolomeo and M. C. Holley. Mechanics of microtubule bundles in pillar cells from the inner ear. *Biophysical Journal*, 73(4):2241–2247, 1997.

J. A. Tolomeo and C. R. Steele. Orthotropic piezoelectric properties of the cochlear outer hair cell wall. *The Journal of the Acoustical Society of America*, 97(5):3006–3011, 1995.

N. Volkmann, D. DeRosier, P. Matsudaira, and D. Hanein. An atomic model of actin filaments cross-linked by fimbrin and its implications for bundle assembly and function. *The Journal of Cell Biology*, 153(5):947–956, 2001.

G. von Békésy. *Experiments in hearing*. Mc Graw Hill, New York, 1st edition, 1960. reprint: American Institute of Physics.

H. L. F. von Helmholtz. Ueber den Einfluss der Reibung in der Luft auf die Schallbewegung. *Verhandlungen des natur-historisch-medizinischen Vereins zu Heidelberg*, 3(17):16–20, 1863.

H. L. F. von Helmholtz. *Die Lehre von den Tonempfindungen als physiologische Grundlage für die Theorie der Musik*. Vieweg, Braunschweig, 5th edition, 1896.

J. R. Womersley. Method for the calculation of velocity, rate of flow and viscous drag in arteries when the pressure gradient is known. *The Journal of Physiology*, 127: 553–563, 1955.

D. E. Zetes. *Mechanical and morphological study of the stereocilia bundle in the mammalian auditory system*. PhD thesis, Stanford University, 1995.

D. E. Zetes and C. R. Steele. Fluid-structure interaction of the stereocilia bundle in relation to mechanotransduction. *The Journal of the Acoustical Society of America*, 101(6):3593–3601, 1997.

O. C. Zienkiewicz, R. L. Taylor, and J. Z. Zhu. *The Finite Element Method: Its Basis & Fundamentals*. Elsevier Butterworth-Heinemann, Oxford, Burlington, 6th edition, 2006.

E. Zwicker. Über die Viskosität der Lymphe im Innenohr des Hausschweines. *Acta Oto-Laryngologica*, 78(1-6):65–72, 1974.

List of Figures

1.1	Micrograph of a hair bundle from the bullfrog's sacculus.	2
1.2	Micrographs of the hair bundles from hair cells of a guinea pig.	3
1.3	Models of the guinea pig inner ear.	6
2.1	Dynamic viscosity measured for Hanks' solution as function of temperature.	17
2.2	Dynamic viscosity measured for Hanks' solution as function of shear rate.	18
2.3	Regions of fluid and solid domain and their boundaries.	27
3.1	Finite element with node locations of the shape functions.	34
3.2	Drag of an oscillating sphere in fluid with the outer boundary free to move.	37
3.3	Drag of an oscillating sphere in fluid with the outer boundary blocked.	38
3.4	Impedance at the inlet of a tube closed at the end and filled with water.	39
3.5	Elastic shell filled with a fluid displaced at the outer boundary in radial direction.	41
4.1	Micrographs of different regions of stereocilia.	44
4.2	Schematic drawing of the hair bundle.	45
4.3	Micrographs of the hair bundle of the bullfrog's sacculus in frontal and side view.	46
4.4	Micrograph of several hair bundles of the bullfrog's sacculus seen from top.	47
4.5	Micrograph of the cuticular plate with the stereocilia and kinocilium removed.	47
4.6	Schematic drawing of the bundle model with geometry parameters . .	48
4.7	Finite-element mesh of the hair bundle model with surrounding liquid.	52
4.8	Mesh of the finite-element hair-bundle model of the stereocilia and the kinocilium.	54
4.9	Mesh of the finite-element hair-bundle model with liquid.	55

5.1	Time traces for a displacement prescribed at the kinocilium and phase relation.	61
5.2	Drag of the hair bundle computed at the kinociliary bulb	64
5.3	Drag of a pivoting cylinder on a fixed plate.	66
5.4	Drag of pivoting cylinders moving in squeezing and shearing modes.	69
5.5	Drag of pivoting cylinders moving in transversal and sliding modes.	69
5.6	Stiffness of the hair bundle computed at the kinociliary bulb.	71
5.7	Length change of tip links divided by the displacement at the kinociliary bulb.	72
5.8	Length change of top connectors divided by the displacement at the kinociliary bulb.	73
5.9	Inertia of the hair bundle computed at the kinociliary bulb.	74
5.10	Time traces for a displacement prescribed at stereocilia on the symmetry plane.	77
5.11	Coherency between stereocilia in the bundle.	78
5.12	Coherency between stereocilia for different top connector stiffness values.	78
5.13	Time traces for a force along all tip links with the same amplitude and phase.	81
5.14	Ratio of the force at the kinociliary bulb divided by the force applied along tip links.	83
5.15	Susceptibility at the short edge in response to forces applied along all tip links.	83
5.16	Displacements at short edge stimulated with a firmly connected probe at the tall edge.	85
5.17	Distortions at short edge stimulated with a firmly or loosely connected probe.	86
5.18	Ratio of the susceptibilities at the short edge.	88
C.1	Variation of the wall-to-wall distance at the tips.	122
C.2	Variation of the tip-link stiffness.	123
C.3	Variation of the top-connector stiffness.	124
C.4	Variation of the kinocilium stiffness.	125
C.5	Variation of the pivotal stereocilia stiffness.	126
C.6	Variation of the viscosity between stereocilia.	127
C.7	Variation of the density of the liquid and all elastic structures.	128

List of Tables

2.1	Order of magnitude estimates of physical quantities for hair bundle mechanics.	21
4.1	Geometrical parameters of the model of the hair bundle from the bullfrog's sacculus.	49
4.2	Material properties of the model of the hair bundle from the bullfrog's sacculus.	57
A.1	Basis μnm units with suitable prefixes for inner ear mechanics.	115
A.2	Derived quantities with prefixes consistent with the μnm units.	116

List of Symbols

Roman Symbols

a	Acceleration
$\mathbf{a}_i$	Orthonormal vector of series expansion
A	Coefficient of series expansion
$\mathbf{B}_u$	Operator defined in (3.21)
B	Coefficient of series expansion
c	Drag coefficient
C	Coefficient of series expansion
$\mathbf{C}$	Damping matrix
c	Speed of sound
c_p	Isobaric specific heat capacity
c_v	Isochoric specific heat capacity
d	Diameter
D	Coefficient of series expansion
$\mathbf{e}$	Vector of strains
$\dot{\mathbf{e}}$	Vector of strain rates
e	Elementary charge
e	Energy
E	Young's modulus
E	Coefficient of series expansion
Ec	Eckert number
EI	Flexural rigidity
f	Frequency
$\vec{f}, F$	Force
$\mathbf{f}$	Vector of external forces
g	Gap distance (wall-to-wall)
$\mathbf{g}$	Vector of external loads
$\mathbf{G}$	Spectral density matrix

h	Height
He	Helmholtz number
i	Imaginary unit
I	Ionic strength
$\mathbf{I}$	Identity tensor
$\mathbf{I_d}$	Deviatoric projection matrix
k	Stiffness
k	Wave number
k_B	Boltzmann constant
k_cof	Coefficient defined in (3.35)
k_th	Thermal conductivity
K	Bulk modulus
$\mathbf{K}$	Stiffness matrix
Kc	Keulegan-Carpenter number
Kt	Number of thermal expansion
l	Length
m	Mass
$\mathbf{m}$	Vector defined in (3.13)
$\mathbf{M}$	Mass matrix
n	Number
$\vec{n}$	Normal vector
N	Basis function
N_A	Avogadro number
$\mathbf{N}$	Matrix of basis functions
NA	Numerical aperture
p	Pressure
$\mathbf{p}$	Vector of pressure degrees of freedom
Pr	Prandtl number
$\vec{q}$	Heat flux
$\dot{q}$	External heat per mass
r	Radius
Re	Reynolds number
s	Spacing
s	Entropy
$\mathcal{S}$	Operator defined in (3.10)
t	Time

t	Sampled time vector
T	Temperature
$\mathbf{T_d}$	Matrix of deviatoric stress
$u, \vec{u}$	Displacement
u	Vector of displacement degrees of freedom
$\vec{v}, v, V$	Velocity
v	Specific volume
w_p	Ansatz function for the pressure
$\vec{w}_u$	Vectorial ansatz function for the displacement
Wo	Womersley number
$\vec{x}$	Space coordinate
x	Cartesian coordinate
y	Cartesian coordinate
z	Cartesian coordinate
Z	Impedance

Greek Symbols

α	Ratio of bulk moduli
β	Coefficient of thermal expansion
Γ	Boundary
γ	Coherency
$\boldsymbol{\varepsilon}$	Strain tensor
$\dot{\boldsymbol{\varepsilon}}$	Strain rate tensor
ε_o	Permittivity of free space
ε_r	Dielectric constant
ε_v	Volume strain
ε_γ	Error of coherency
ζ	Damping ratio
η	Dynamic viscosity
η^B	Bulk viscosity
λ	Isotropic material coefficient
λ'	Isotropic material coefficient
μ	Isotropic material coefficient

μ'	Isotropic material coefficient
ν	Poisson's ratio
ξ	Coefficient in (5.9)
ϱ	Density
σ	Coefficient in (5.7)
$\boldsymbol{\sigma}$	Stress tensor
$\boldsymbol{\tau}$	Deviatoric stress tensor
ϕ	Angle between the bundle sides with the origin at the kinocilium
Φ	Dissipation function
χ	Coefficient in (5.9)
χ	Susceptibility
$\boldsymbol{\chi}$	Susceptibility matrix
ω	Circular frequency
Ω	Region

Others

$(\cdot)^*$	Dimensionless value
$\tilde{(\cdot)}$	Experimental quantity
$\hat{(\cdot)}$	Amplitude of complex value
$\dot{(\cdot)}$	First time derivative
$\ddot{(\cdot)}$	Second time derivative
$(\cdot)^\mathsf{T}$	Transpose
∇	Nabla operator
$\mathrm{tr}(\cdot)$	Trace operator
$\mathrm{Re}(\cdot)$	Real-part operator
$\mathrm{Im}(\cdot)$	Imaginary-part operator
$\mathbf{0}$	Zero matrix

Subscripts

$(\cdot)_\mathrm{b}$	Bottom
$(\cdot)_\mathrm{bl}$	Boundary layer
$(\cdot)_\mathrm{cyl}$	Cylinder

$(\cdot)_{dic}$	Differential interference contrast microscopy
$(\cdot)_{dis}$	Distortion
$(\cdot)_{edg}$	Edge
$(\cdot)_{ell}$	Ellipsoid
$(\cdot)_{fluid}$	Fluid
$(\cdot)_{fsi}$	Fluid-structure interaction
$(\cdot)_{hb}$	Hair bundle
$(\cdot)_{tc}$	Top connector
$(\cdot)_{in}$	Internal
$(\cdot)_{kc}$	Kinocilium
$(\cdot)_{kb}$	Kinociliary bulb
$(\cdot)_{kl}$	Kinociliary link
$(\cdot)_m$	Mean
$(\cdot)_o$	Origin
$(\cdot)_p$	Pressure
$(\cdot)_r$	Rootlet
$(\cdot)_{ref}$	Reference
$(\cdot)_{rel}$	Relative
$(\cdot)_{rot}$	Rotational, pivotal
$(\cdot)_{solid}$	Solid
$(\cdot)_{sph}$	Sphere
$(\cdot)_{sc}$	Stereocilium
$(\cdot)_t$	Tip
$(\cdot)_{tl}$	Tip link
$(\cdot)_{tot}$	Total
$(\cdot)_{th}$	Thermal
$(\cdot)_u$	Displacement
$(\cdot)_w$	Properties corrected for boundary-layer effects

Appendix A

Units

Physical quantities of the mechanics of the inner ear are on a small scale for the geometry and fast in time compared to classical engineering problems. With respect to the typical dimensions the micrometer turns out to be a suitable length. For hearing one kilohertz is representative for audible frequencies. The inverse provides the timescale by the millisecond. As the pressure is usually in the order of one Pascal, the appropriate mass is in nanogram. The electrical potentials are in the range of millivolts and thus the electrical current should be given in picoampere.

The base units of the international system SI (2006) with the appropriate prefixes are listed in Tab. A.1.

Table A.1: Basis μnm units with suitable prefixes for inner ear mechanics.

Name	Units with prefix
Length	μm
Mass	ng
Time	ms
Electric current	pA
Thermodynamic temperature	K
Amount of substance	mmol

For convenience the derived quantities should have consistent prefixes. Some examples of derived quantities commonly used here are given in Tab. A.2. To increase the readability some of the prefixes are rearranged. In the style of the abbreviation mks (meter-kilogram-second) for the standard SI units and the cgs (centimeter-gram-second) here the abbreviation μnm (micrometer-nanogram-millisecond) is used. As far as possible these units are used throughout this work.

Table A.2: Derived quantities with prefixes consistent with the μnm units.

Quantity	Units with prefix
Density	$Gg \cdot m^{-3}$
Drag	$nN \cdot s \cdot m^{-1}$
Energy	aJ
Frequency	kHz
Force	pN
Pressure	Pa
Power	fW
Stiffness	$\mu N \cdot m^{-1}$
Voltage	mV

Appendix B

Drag Coefficients of Two Closely Apposed Cylinders

B.1 Pivoting to their Common Center

The Stefan-Reynolds equation describes the pressure p in a small gap of a viscous and incompressible liquid for moving walls (Nguyen 2000). Based on this equation the drag between two closely apposed cylinders pivoting towards their common center is approximated. For the derivation the software Mathematica (Version 7.0, Wolfram Research Inc., Champaign, Illinois) was used.

The squeeze flow between two disks moving to their common center is the basis for the derivation. For a gap distance depending only on one spatial variable the relation reads

$$\frac{\partial}{\partial x}\left(g^3 \frac{\partial p}{\partial x}\right) = 12\eta \frac{\partial g}{\partial t}. \tag{B.1}$$

with the dynamic viscosity η, the time t, and the gap distance g. The direction x is normal to the line connecting the centers of the disks and the origin of the coordinate is located at their common center. The geometry of the gap height, the wall-to-wall distance between two circular disks with the same radius r is

$$g(x) = g_0 + 2r\left(1 - \sqrt{1 - \left(\frac{x}{r}\right)^2}\right) \tag{B.2}$$

with the minimal wall-to-wall distance g_0. A second-order Taylor-series expansion at the origin provides an approximation of the geometry as

$$g(x) \approx g_0 + \frac{x^2}{r}. \tag{B.3}$$

Both disks are moving towards their common center with the speed $v/2$. The relative velocity between them is v and thus the time derivative of the gap height

$$\frac{\partial g}{\partial t} = -v \tag{B.4}$$

is independent of x and negative as the disks are moving towards each other.

The solution of the differential equation for the pressure (B.1) with the gap height relations (B.3) and (B.4) reads

$$p(x) = \frac{3 r^3 v \eta}{(g_0 r + x^2)^2}$$
$$+ C_1 \left(\frac{3x}{8 g_0^2 r^2 (g_0 r + x^2)} + \frac{3 \arctan\left(\frac{x}{\sqrt{g_0 r}}\right)}{8 g_0^{5/2} r^{5/2}} + \frac{x}{4 g_0 r (g_0 r + x^2)^2} \right) + C_2 . \tag{B.5}$$

The two unknown integration constants C_1 and C_2 are determined from two boundary conditions. One condition is that at infinity the pressure should be zero

$$\lim_{x \to \infty} p(x) = 0 \tag{B.6}$$

The second is the symmetry constrain at the origin

$$\left. \frac{\partial p(x)}{\partial x} \right|_{x=0} = 0 \tag{B.7}$$

Herewith the pressure equation (B.5) simplifies to

$$p(x) = \frac{3 r^3 v \eta}{(g_0 r + x^2)^2} , \tag{B.8}$$

The force which acts on the disks is the integral of the pressure over x from $-\infty$ to ∞. The evaluated integral reads

$$c_{\text{disk}} = \frac{3}{2} \pi \left(\frac{r}{g_0} \right)^{3/2} \eta . \tag{B.9}$$

where c is the drag coefficient defined as the ratio of force divided by velocity. This is the same relation as derived by Zetes (1995).

To compute the drag of the pivoting cylinders in their slightly oblique position, it is assumed that the flow is essentially two-dimensional and the effective drag coefficient can be computed by integrating over the height. The minimal wall-to-wall distance is a function of the height coordinate z as

$$g_o(z) = g_b - \frac{g_b - g_t}{h} z \qquad (B.10)$$

with the wall-to-wall distance at the bottom g_b and at the tip g_t. The height of the cylinders is h. For the pivotal motion the relative velocity is also a function of the height as

$$v(z) = v_t \frac{z}{h} \qquad (B.11)$$

where v_t denotes the velocity at the cylinder tip. To satisfy the balance of momentum with respect to the center of rotation the equivalent force at the tip

$$F_t = \frac{1}{h} \int_{z=0}^{h} c_{\text{disk}}(z) \, v(z) \, z \, dz \, . \qquad (B.12)$$

has to counteract the pressure forces of c_{disk} on the right-hand side. The force at the height h is denoted F_t. Finally the solution of the integral yields the drag coefficient for two apposed cylinders pivoting towards their common center as

$$c_{\text{squeeze}} = \pi \eta h \frac{\xi^2 (3+\xi) \chi^3}{(1+\xi)^3} \quad \text{with} \quad \xi = \sqrt{\frac{g_t}{g_b}} \quad \text{and} \quad \chi = \sqrt{\frac{r}{g_t}} \, . \qquad (B.13)$$

The drag coefficient represents the force needed at the tips to squeeze the cylinders together for a given velocity.

B.2 Pivoting in the Same Direction

The drag associated with the pivotal motion of two cylinders moving into the same direction and in the plane containing their axes is estimated here. The drag is induced by the shear of the fluid in between the two cylinders. If the common translatory motion is removed, the only motion left that induces drag is their relative motion as both cylinders move along their axis but in opposite directions. The analytical solution by Hunt et al. (1994) provides an analytical expression for the drag of a cylinder moving along its axis parallel to a plane wall, which is equivalent to two parallel cylinders moving in opposite directions along their axes. The drag coefficient per length for this mode of motion is

$$c_{\text{slice}} = \frac{2 \pi \eta}{\text{arccosh}\,(g_o/r + 1)} \qquad (B.14)$$

based on the relative velocity between the two cylinders with the dynamic viscosity η, the radius of the cylinders r, and the wall-to-wall distance between the cylinders g_o. Using the same approach as in the previous section with a place dependent gap distance allows to compute the drag for the slightly oblique arrangement. The integral to solve reads

$$F_{\text{rel}} = v_{\text{rel}} \int_{z=0}^{h} c_{\text{slice}}(z) \, dz \ . \tag{B.15}$$

where the height dependency of c_{slice} enters by the wall-to-wall distance relation from Equation (B.10). For this integral no closed solution was found. The integral was evaluated numerically with Mathematica (Version 7.0, Wolfram Research Inc., Champaign, Illinois).

The kinematic relation of the pivotal motion relates the relative velocity between the cylinders v_{rel} to the velocity at the tip v_t by

$$v_{\text{rel}} = v_t \frac{r}{h} \ . \tag{B.16}$$

The equivalent force along the direction of the cylinder F_{rel} scales by this lever arm ratio as

$$F_{\text{rel}} = F_t \frac{h}{r} \ . \tag{B.17}$$

Thus the drag coefficient for forces and velocities along the cylindrical axes translates back into the coefficient with forces and velocities at the tips of the pivoting cylinders by the ratio

$$\frac{F_t}{v_t} = \frac{r^2}{h^2} \frac{F_{\text{rel}}}{v_{\text{rel}}} \ . \tag{B.18}$$

This provides an estimate of the drag coefficient related to the sliding motion between pivotal cylinders.

Appendix C
Parameter Dependence of the Bundle Model

C.1 Wall-to-Wall Distance at the Tips

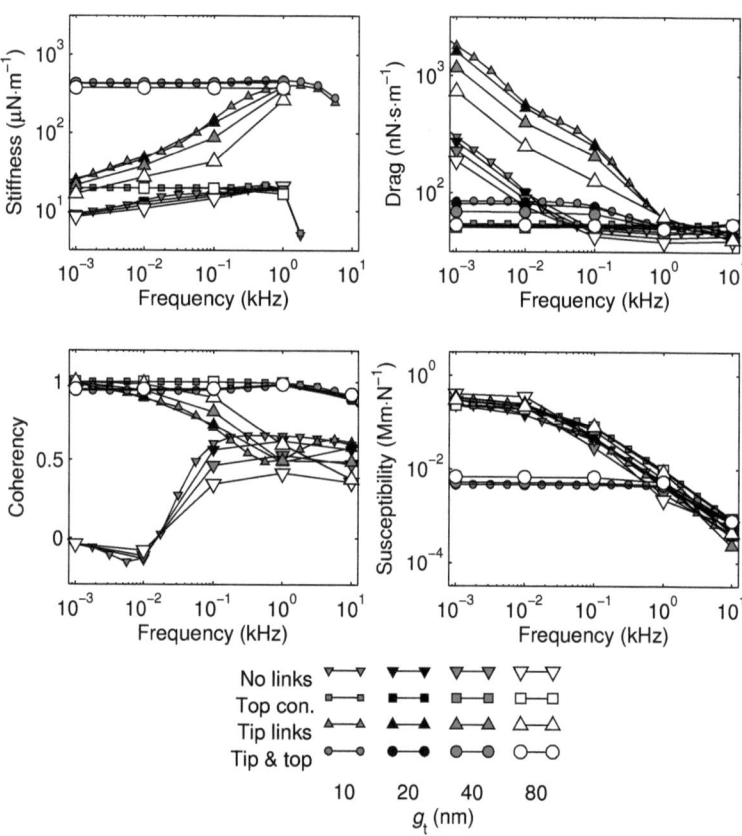

Figure C.1: Variation of the wall-to-wall distance at the tips. The parameter variations are indicated by different marker sizes and fillings. The used values are given below all plots. The reference values have a shorter marker spacing over the frequency. They are discussed in Section 5.1 for the stiffness and the drag, in Section 5.2 for the coherency, and in Section 5.3 for the susceptibility.

C.2 Tip-Link Stiffness

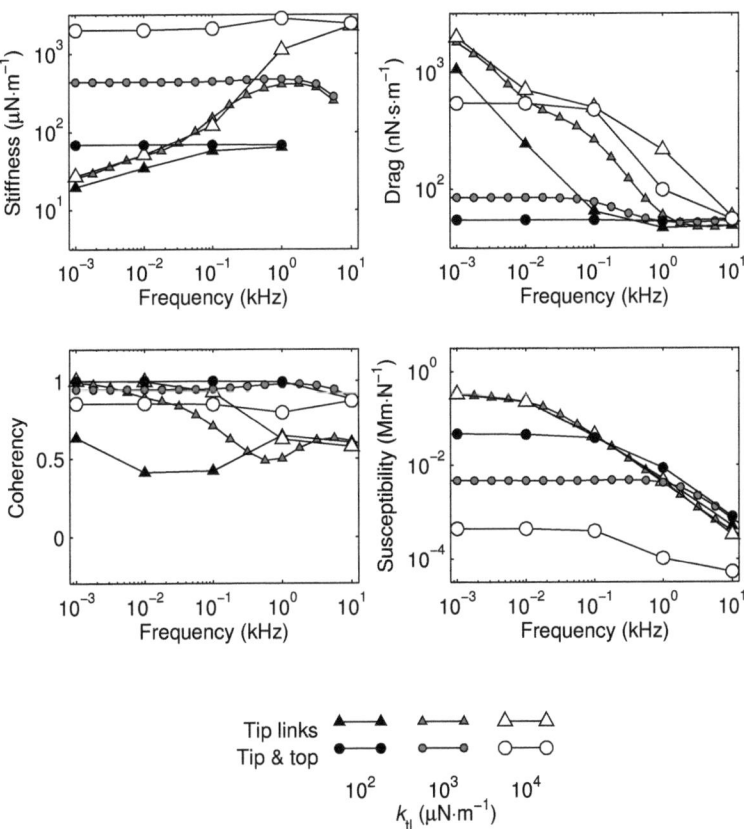

Figure C.2: Variation of the tip-link stiffness. The parameter variations are indicated by different marker sizes and fillings. The used values are given below all plots. The reference values have a shorter marker spacing over the frequency. They are discussed in Section 5.1 for the stiffness and the drag, in Section 5.2 for the coherency, and in Section 5.3 for the susceptibility.

C.3 Top-Connector Stiffness

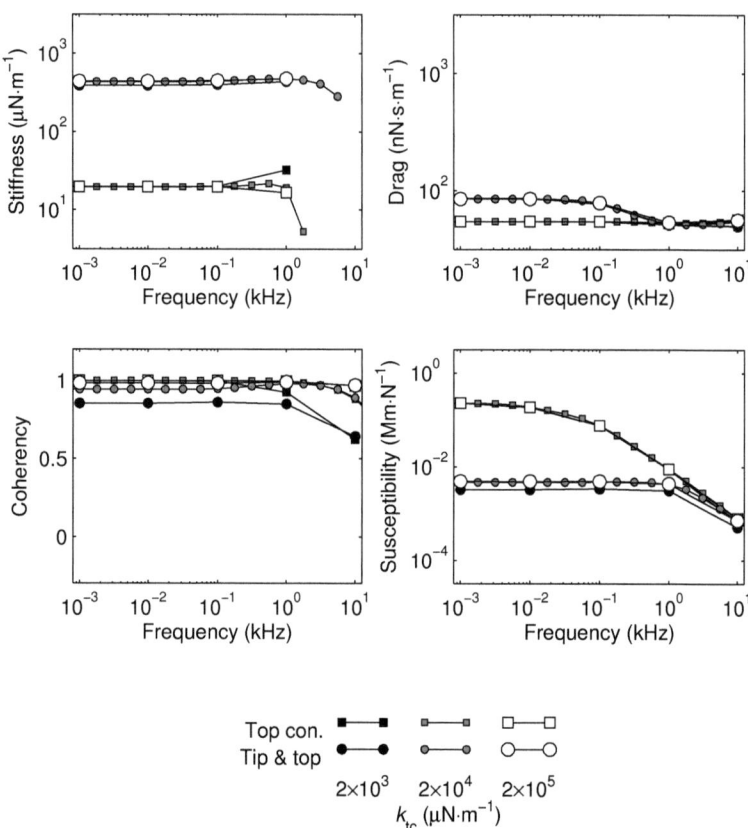

Figure C.3: Variation of the top-connector stiffness. The parameter variations are indicated by different marker sizes and fillings. The used values are given below all plots. The reference values have a shorter marker spacing over the frequency. They are discussed in Section 5.1 for the stiffness and the drag, in Section 5.2 for the coherency, and in Section 5.3 for the susceptibility.

C.4 Kinocilium Stiffness

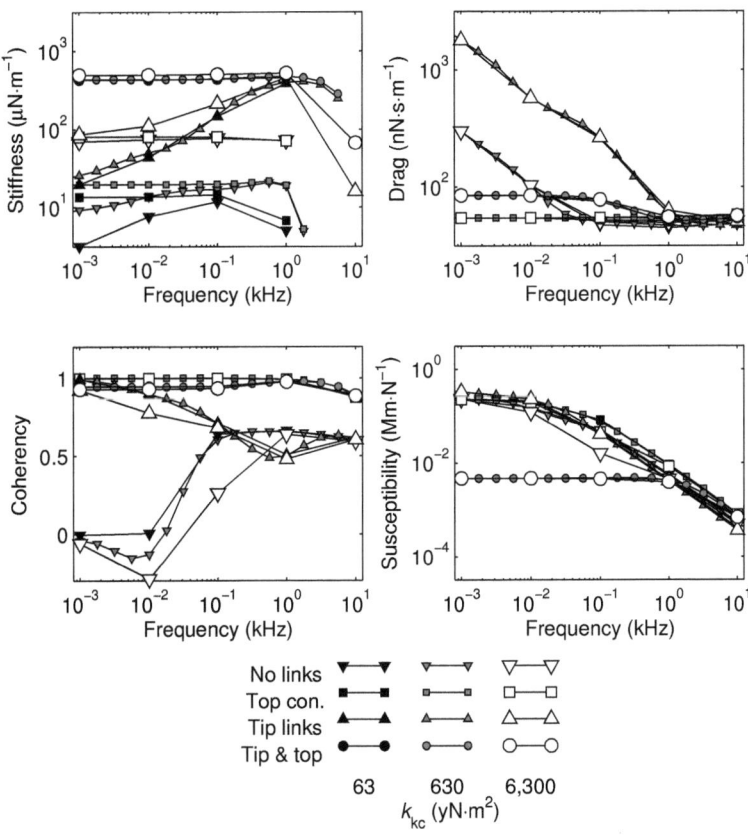

Figure C.4: Variation of the kinocilium stiffness. The parameter variations are indicated by different marker sizes and fillings. The used values are given below all plots. The reference values have a shorter marker spacing over the frequency. They are discussed in Section 5.1 for the stiffness and the drag, in Section 5.2 for the coherency, and in Section 5.3 for the susceptibility.

C.5 Pivotal Stereocilia Stiffness

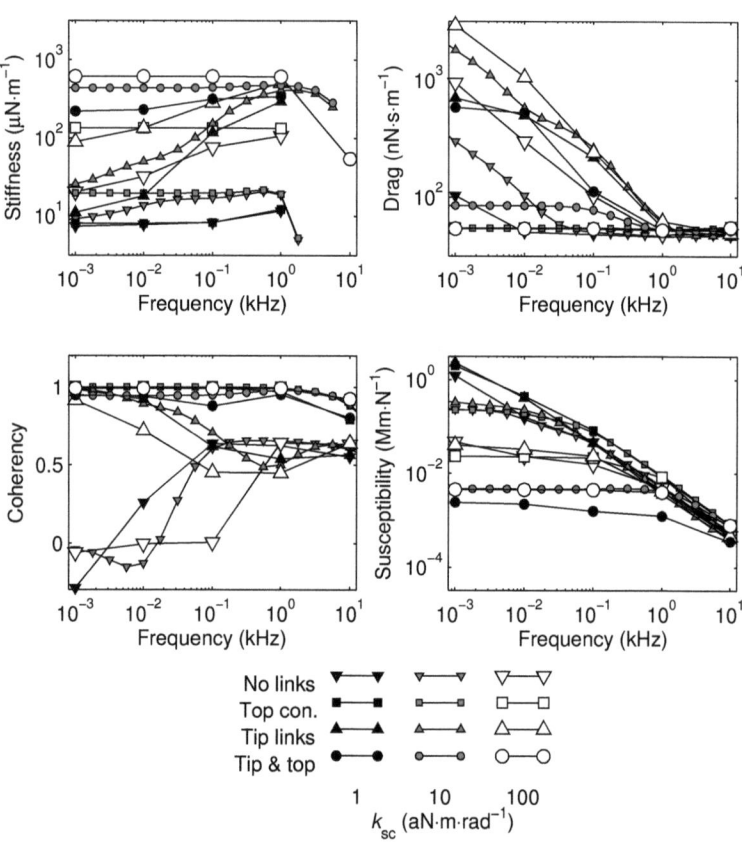

Figure C.5: Variation of the pivotal stereocilia stiffness. The parameter variations are indicated by different marker sizes and fillings. The used values are given below all plots. The reference values have a shorter marker spacing over the frequency. They are discussed in Section 5.1 for the stiffness and the drag, in Section 5.2 for the coherency, and in Section 5.3 for the susceptibility.

C.6 Viscosity between Stereocilia

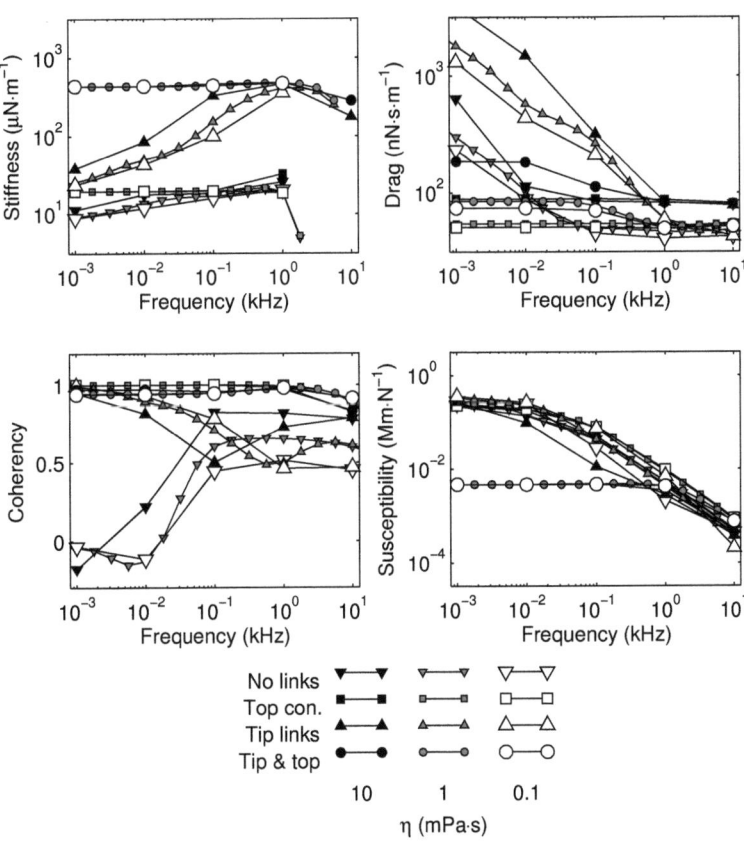

Figure C.6: Variation of the viscosity between stereocilia from the tips down by 0.325 µm. The parameter variations are indicated by different marker sizes and fillings. The used values are given below all plots. The reference values have a shorter marker spacing over the frequency. They are discussed in Section 5.1 for the stiffness and the drag, in Section 5.2 for the coherency, and in Section 5.3 for the susceptibility.

C.7 Density

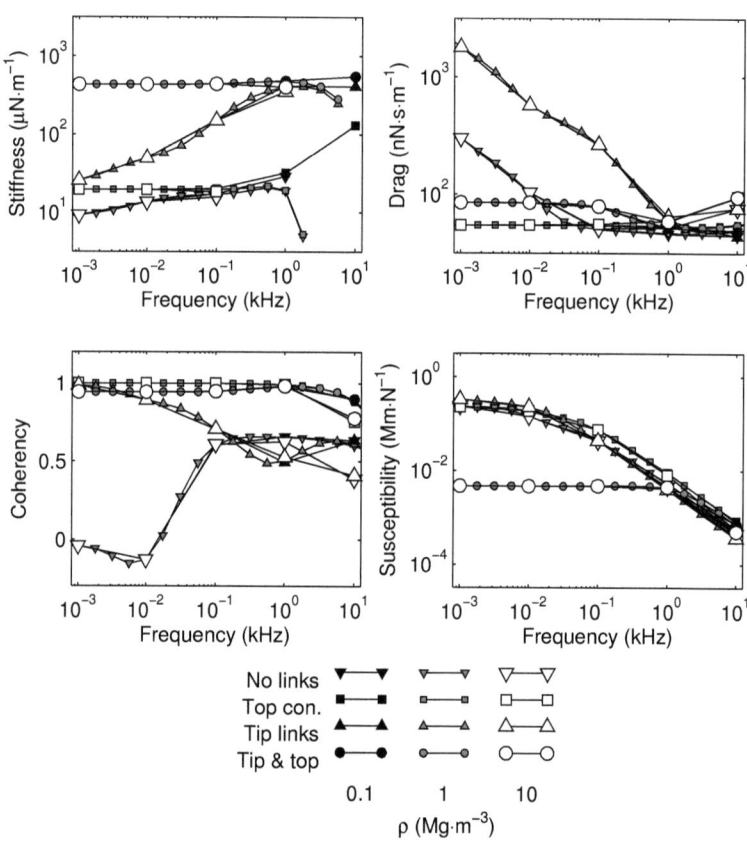

Figure C.7: Variation of the density of the liquid and all elastic structures. The parameter variations are indicated by different marker sizes and fillings. The used values are given below all plots. The reference values have a shorter marker spacing over the frequency. They are discussed in Section 5.1 for the stiffness and the drag, in Section 5.2 for the coherency, and in Section 5.3 for the susceptibility.

Die VDM Verlagsservicegesellschaft sucht für wissenschaftliche Verlage abgeschlossene und herausragende

Dissertationen, Habilitationen, Diplomarbeiten, Master Theses, Magisterarbeiten usw.

für die kostenlose Publikation als Fachbuch.

Sie verfügen über eine Arbeit, die hohen inhaltlichen und formalen Ansprüchen genügt, und haben Interesse an einer honorarvergüteten Publikation?

Dann senden Sie bitte erste Informationen über sich und Ihre Arbeit per Email an *info@vdm-vsg.de*.

Sie erhalten kurzfristig unser Feedback!

VDM Verlagsservicegesellschaft mbH
Dudweiler Landstr. 99　　　　　　Telefon　+49 681 3720 174
D - 66123 Saarbrücken　　　　　　Fax　　　+49 681 3720 1749
www.vdm-vsg.de

Die VDM Verlagsservicegesellschaft mbH vertritt

Printed by Books on Demand GmbH, Norderstedt / Germany